Earth Structures Engineering

Earth Structures Engineering

R.J. Mitchell

Queen's University, Kingston, Ontario

Boston
ALLEN & UNWIN, INC.
London   Sydney

Allen & Unwin Inc.,
9 Winchester Terrace, Winchester, Mass. 01890, USA

George Allen & Unwin (Publishers) Ltd,
40 Museum Street, London WC1A 1LU, UK

George Allen & Unwin (Publishers) Ltd,
Park Lane, Hemel Hempstead, Herts HP2 4TE, UK

George Allen & Unwin Australia Pty Ltd,
8 Napier Street, North Sydney, NSW 2060, Australia

First published in 1983

Library of Congress Cataloging in Publication Data

Mitchell, Robert J.
  Earth structures engineering.
Includes bibliographical references and index.
1. Earthwork.   2. Soil mechanics.   I. Title.
TA715.M53   1983   624.1'891   83-2621
ISBN 0-04-624003-9
ISBN 0-04-624004-7 (pbk.)

British Library Cataloguing in Publication Data

Mitchell, Robert J.
  Earth structures engineering.
1. Earthwork
I. Title
624.1'6   TA715
ISBN 0-04-624003-9
ISBN 0-04-624004-7 Pbk

Set in 10 on 11 point Times Roman by Bedford Typesetters Ltd,
and printed in Great Britain
by Mackays of Chatham

Preface

Earth structures engineering involves the analysis, design and construction of structures, such as slopes and dams, that are composed mainly of earth materials, and this is a growth area in geotechnical engineering practice. This growth is due largely to increased involvement in designing various types of earth structures for the resources industries (slopes, impoundment structures, offshore islands, mine backfills), to the development of increasingly large hydroelectric projects, to the need for more freshwater storage and diversion schemes, and to the need for transportation, communications and other facilities in areas where the natural earth materials are occasionally subject to mass instabilities.

Although geotechnical engineering transects traditional disciplinary boundaries of civil, geological and mining engineering, the majority of geotechnical engineers are graduates from civil engineering schools. Here the geotechnical instruction has been concentrated on soil mechanics and foundation engineering because foundation engineering has traditionally been the major component of geotechnical practice. Geotechnical specialists, however, generally have acquired considerable formal or informal training beyond their first engineering degree, and an advanced degree with considerable cross-discipline course content is still considered an advantage for a young engineer entering a career in geotechnical engineering. Practical job experience is, of course, a necessary part of professional development but is readily interpreted and assimilated only if the required background training has been obtained. In recent years many engineering schools have, for a variety of reasons (including improved computational abilities and the changing role of professional engineers with the availability of more technologists) found room in their programs for more technical options (hence greater specialization) at the undergraduate level. In the geotechnical area, options in various aspects of earth structures engineering are an obvious choice and such course offerings can be found in many university calendars. Although texts, detailed design manuals, research papers and state-of-the-art reports have been published in most areas of geotechnical engineering, earth structures engineering is one area where there does not appear to be a publication of sufficient scope to serve as a basis for instruction. It is the purpose of this text to present the basic concepts, principles and procedures used in earth structures engineering at a senior undergraduate to graduate university level, and it is assumed that the reader has a basic knowledge of earth materials (basic geology) and earth mechanics (basic soil and rock mechanics).

Chapters 1 and 2 are to some extent a review of concepts, material properties and analyses detailed in basic geotechnical courses, but are also designed to indicate some differences between the practices of foundation design and earth structures design and to extend some of the concepts and methodology particularly relevant to earth structures engineering. For example: the structural integrity and performance of a large earth dam will depend on the correct evaluation of the properties and variability of natural materials over an extensive area and suitable earth materials must be located to construct a safe yet economical structure – this additional emphasis on geological engineering is reflected in the brief reviews of terrain analysis and site investigation techniques presented in Chapters 1 and 2; earth structures are designed on limiting equilibrium (rather than limiting settlements) and ground water is a major design consideration (rather than a nuisance factor), yet design safety factors are, for reasons of economic viability, relatively low (1.5 rather than 3.0) and failures are relatively catastrophic – this additional emphasis on a more detailed interpretation of the shear behavior of earth materials is reflected in the brief reviews of strength and groundwater flow in Chapter 2. Chapters 1, 2 and 3 combine to provide basic design and analysis information for engineers concerned with the geotechnical aspects of transportation facilities and route selection.

Slope stability problems affect all types of land developments, from single dwelling residential buildings on ravine lots to major projects such as canals and reservoirs – Chapter 4 details the basic analyses and design considerations applicable to slope stability problems. Earth dams are continually being constructed and maintained for hydroelectric power production, water storage, recreation and waste management – Chapter 5 outlines the basic design concepts, principles and requirements for earth dams, discusses detailed design and construction considerations for typical situations and summarizes the state-of-the-art in dam design. Case study and research references are provided for more advanced study. Ground subsidence and mine backfill are discussed in Chapter 6. Example problems and tutorial problems are presented in Chapters 4, 5 and 6.

Although the book is specifically intended as a teaching text in civil and geological programs, practising geotechnical engineers should find it useful as a library reference. Mining engineers will also find analyses and design information and examples pertinent to mine operations throughout the book. After 15 years of research, consulting and teaching contact with engineers and students in civil, geological and mining engineering, the author has selected and presented material to form a teaching and reference text for today's geotechnical option students, tomorrow's earth structures engineers.

Robert J. Mitchell
Kingston, Ontario
JUNE 1982

Acknowledgements

It is not possible to acknowledge all of the many associates, friends and former teachers who have contributed to the development of this text, but special thanks are extended to my family for their support and encouragement. Credits are also due to Mrs Hazel Walker for typing the draft manuscripts and to Mr Jim Roettger for assisting in the preparation of figures.

The following aerial photographs in Chapter 1 are copyrighted between 1929 and 1980 to Her Majesty the Queen in Right of Canada and are reproduced with permission of Energy, Mines and Resources, Canada: 1.4a and b, 1.7, 1.8, 1.9, 1.10b and c, 1.12b and c, 1.13a, b and c, 1.14, 1.15, 1.16a and b, 1.17, 1.19b, 1.20a and b, 1.21a and b, 1.24, 1.25, 1.26 and 1.28. References to other illustrations and information sources appear throughout the text, and photo credits are noted in the captions.

Contents

Preface	*page*	vii
Acknowledgements		ix
List of tables		xiii

1   Earth structures and air photo interpretation

1.1	Earth materials: soils and rock	1
1.2	Basic information sources	2
1.3	Air photo interpretation	4
1.4	Rock landforms	12
1.5	Glacial landforms	12
1.6	Lacustrine, marine and alluvial landforms	16
1.7	Eolian and residual landforms	18
1.8	Uses of air photos in earth structures engineering	19
1.9	Other remote sensing and probing techniques	27
1.10	Problems on air photo interpretation	33

2   Earth mechanics in earth structures engineering

2.1	Strength and deformation of earth materials	39
2.2	Ground water and earth structures	58
2.3	Settlement of earth structures	74

3   Embankments and tunnels

3.1	Embankments on soft ground	87
3.2	Soft-ground tunneling	96
3.3	Problems on bearing capacity and tunnels	101

4   Slope stability

4.1	Types of slope movements	104
4.2	Slope stability analyses	109
4.3	Design charts for slopes in homogeneous materials	125
4.4	Crest loadings, dynamic loadings, submergence and drawdown	135
4.5	Recommended design factors of safety and procedures	142
4.6	Construction considerations and remedial measures	145
4.7	Permanent retaining walls	151
4.8	Problems on slope stability	154

5 Earth dams

5.1	Types of earth dams	*page* 163
5.2	Dam design considerations	163
5.3	Foundation treatments and efficiencies	180
5.4	Dam settlements and distortion	190
5.5	Earthquake and rapid drawdown design	198
5.6	Some special considerations in construction of earth dams	199·
5.7	Monitoring, performance and maintenance of earth dams	201
5.8	Mine-tailings dams and process-water impoundments	205
5.9	Problems on earth dams	214

6 Ground subsidence and mine backfill

6.1	Ground control using backfill	219
6.2	Cemented tailings backfill design	226
6.3	Use of uncemented tailings backfill	232
6.4	Subsidence and surface effects	236
6.5	Problems on mine backfill and subsidence	241
	Appendix: units and symbols	245
	References	247
	Answers to problems	261
	Index	264

List of tables

1.1	Features and engineering use of common Pleistocene landforms	*page* 16
2.1	Brief summary of drilling and sampling methods	39
2.2	Engineering classification of earth materials	40
2.3	Brief summary of common field tests	41
2.4	Brief summary of common laboratory tests	42
2.5	Effects of σ_2' on soil strength	51
2.6	Strength anisotropy	52
2.7	Typical permeabilities of earth materials	59
2.8	Typical properties of compacted materials	77
2.9	Typical strength and compression characteristics of earth materials	79
4.1	Stability numbers for temporary slopes in homogeneous soils	134
4.2	Typical safety factors	143
4.3	Summary of slope stability design and construction	144
4.4	Some case studies of slope stability problems	145
5.1	Cutoffs for dam foundations	181
5.2	Treatment of foundation soils	182
5.3	Dam settlement and cracking	198
5.4	Compaction of core materials	202
5.5	Some case studies of dam and reservoir failures	206
5.6	Some dams requiring special construction or remedial measures	208
5.7	Mine waste disposal considerations	210
6.1	Mine waste backfills	221

1 Earth structures and air photo interpretation

In earth structures engineering, the recognition of potential problems is often a greater challenge than the actual design. Many problems are regional in scope and considerable preliminary work must preccde an effective and efficient site investigation. Of fundamental concern are the occurrence and distribution of material types and ground water; the interpretation of air photos can provide preliminary information of this type for problem recognition and site planning purposes. Air photos and infrared scanner images are also used for agriculture, waste disposal, resource development and many other planning activities.

This chapter provides an introduction to air photo recognition elements pertinent to earth structures engineering and briefly discusses the use of maps and air photos in slope stability and earth dam site investigations. Further study of detailed interpretation manuals (for example, Avery 1968, Mollard 1973, Way 1973) is recommended for development of interpretative skills.

1.1 Earth materials: soil and rock

The solid portion of the Earth's crust is composed of soil and rock. These earth materials include water and organic materials that might exist together with the basic minerals which form the soil or rock. The major engineering distinction between soils and rocks is that soils can be separated into individual particles by gentle mechanical agitation in water while rocks are held together by strong primary bonding. Geologists often refer to soils as unconsolidated materials and mining engineers refer to cemented sands, used in mine backfill, as being consolidated. The geotechnical engineer must become familiar with this usage while the geologists and mining engineers must appreciate that consolidation, to a geotechnical engineer, is the hydrodynamic process by which water is forced out of a soil under applied loads. There are definite relations, however, between the types of material landforms naturally occurring in the Earth's crust and the geological processes by which these materials were formed – consolidation is one of these processes. A basic understanding of geological processes is a necessary background to

1

the study of air photos and to the successful practice of earth structures engineering.

1.2 Basic information sources

Prior to this century all maps were prepared from ground survey data. Geologists often traversed remote wilderness areas to produce the first detailed maps of these regions. Today, vertical air photos are taken from aircraft flying at a fixed altitude and fixed ground speed along prescribed parallel flight lines to produce terrain photo images from which topographic maps are plotted. The science of producing accurate maps from air photos is called photogrammetry and the tools have developed, during the last two or three decades, from the early simple type of plotting machine pictured on Figure 1.1, to modern stecometers where the operator can put co-ordinate data from air photos onto a computer which corrects (using vertical and horizontal ground control data previously input) for photo tilt and distortion and then automatically maps the information. Indeed, modern technology can produce a great variety of maps including orthophoto maps printed directly from air photo controlled mosaics with superimposed contours and other information.

Topographic maps are available from government agencies in most countries and the land masses are generally mapped at a scale of 1:250000 showing natural terrain, major developments and providing 100 ft (30.5 m) contour intervals. Most populated land areas are mapped at a scale of 1:50000 which provides additional detail (pits, quarries, buildings, cemeteries, bridges, marshland, depressions, forested areas, dumps, road information) and 25 ft (7.6 m) contours. Heavily populated areas may be mapped at a scale of 1:25000 or less and may provide 5 ft (1.5 m) contours plus some additional cultural or land use information. It is not possible, however, to put all the information contained on the photo image onto a map. Features such as landforms, drainage details and natural slope stability are left for photo interpretation. In addition, maps are often outdated in urban areas while recent photographs may be available. While national agencies are the main source of air photos, there may be state or provincial air photo libraries, and some larger regional or municipal governments have their own photos (as well as maps produced from these photos). Photo prints can usually be purchased from government libraries at a nominal cost for reproduction.

Bedrock geology maps prepared from air photos (outcrops are easily identified and fault zones may often be inferred) and well drilling records, supplemented by core drilling and recent techniques such as radar imagery, are also available from government agencies. These are most valuable for site selection and for planning site investigations. Physiographic soils maps

2

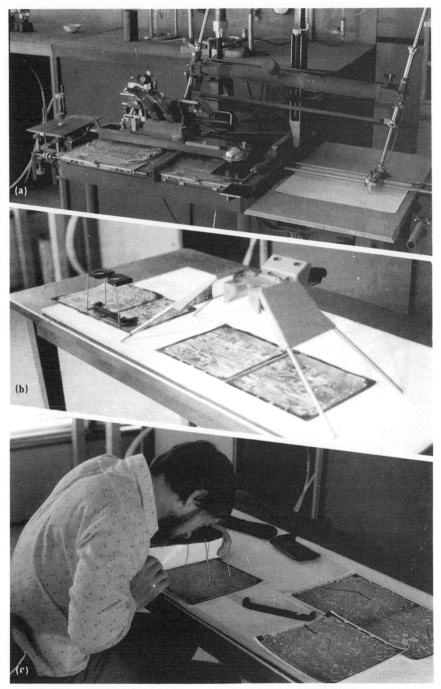

Figure 1.1   Air photo interpretation methods. (a) Early stereo plotting equipment. (b) Pocket glasses, parallax bar and mirror stereoscope. (c) Use of photographs and pocket glasses.

are generally published by state or provincial agencies and are used mainly for agricultural and forestry purposes. Surficial soils information can be obtained from these maps, however, making them a useful reference for air photo interpretation studies. The data storage and retrieval capacity of modern computers has led to new information services being available from many government agencies and private firms often contribute to the database. For example, the Ontario Ministry of Transportation and Communications (MTC) have field data from bridge site investigations available on a computer system. The practicing geotechnical engineer should be aware of all sources of information on local soil properties and stratigraphy.

In remote areas where detailed maps are not available, air photos provide the basic source of information for terrain analysis. Sometimes large-scale (1 : 5000) photos of particular sites (or routes) may be desired after initial study of the commonly available 1 : 15000 photos. This can be done, as can colour photography or infrared scanning, by private aviation companies but the cost per photo will be substantial, the major cost being the flying time. Nevertheless, many mining companies and other developers do have photos flown for planning and mapping purposes.

Materials and groundwater information from photo interpretations can be added to base maps (topographic and geological) for the purposes of preliminary material quantity estimates and locating field site studies.

1.3   Air photo interpretation

The air photo interpreter is a geodetective using clues from the photos together with his background knowledge and experience to provide a framework for site investigations. Photos are sequenced, along a flight line, to provide about 60% overlap such that two sequentially numbered photographs can be placed side-by-side under a mirror stereoscope to obtain a stereo view of the overlapped area. Every second photo along a flight line may be laid out to form a linear map with the intermediate photos being used to obtain stereo images of the landscape (see Fig. 1.1). Pocket stereo glasses may be used in place of the more expensive mirror stereoscope and the standard 230 mm square photos can be rolled to obtain a stereo image of various landforms. There is no need to cut photos to use pocket stereo glasses, and careful rolling or curling up of the edges to provide stereo viewing of the middle area will not damage the photos. Stereo pairs published in texts and manuals are made by cutting the same area from sequential photos and fitting these for viewing with stereo glasses. Flight lines are usually drawn on 1 : 250000 topographic maps and such index maps are available at photo libraries. Air photos are generally designated by a film roll number followed by an individual photo number and these numbers are noted on the flight lines. Index maps, and the first photo of each roll, provide information on the time and date of the photos, photo scale, film type, camera focal length and flying height. A stereo image of part of the terrain may also be obtained across flight lines where there is about 30% overlap, but there is usually more distortion between such combinations than between photos taken along the same line.

The first thing the amateur interpreter will recognize is the pronounced vertical exaggeration in the stereoscopic image – buildings look to be much higher than they really are. This is due to parallax, defined as the apparent displacement of a point in a fixed reference system as a result of a change in the point of observation; the photos are taken several kilometers apart along a flight line, giving two different perspectives on each object. When these different perspectives are fused together under stereoscopic viewing, the difference in parallax between the top and bottom of the object will appear as an exaggerated difference in altitude. It is this phenomenon that allows accurate height (or depth) measurements and accurate surface contouring to be accomplished, since the operator of a stereoscopic plotting machine is simply correcting for parallax by floating two reference points which are actually fixed into the viewing telescopes.

The second interesting observation may be that, when the photo positions are reversed, open wells will look like silos and silos will look like wells, all the houses will go underground and rivers will run on the tops of levees. This is reversed parallax – to see the correct image (or terrain model) the photos must be correctly orientated. Amateur observations are likely to be most rewarding: a winding elevated roadway seen on a topographic map is actually located on an esker; local cemeteries on the topographic map are located in ancient beach sand deposits; there is evidence of artesian groundwater pressures and quicksand in the marshland seen on a topographic map; it may be that a rock fault underlies a marked linear feature seen on a photo (or is it an ancient shoreline?); sand borrow might be obtained from the pine-forested kame terrace not noted on the topographic map; the subsurface farm drainage tiles, seen as fine white lines on photos, will have to be dug up before the proposed lagoon is constructed or leakage will be a problem; why was a development placed so close to an old landslide scar where tilted trees indicate that slope movements are still occurring and the abundance of cedar trees in the gullies indicates high groundwater seepage? A great number of interesting observations can be made immediately from air photos.

It takes little skill, and only modest geological background and geotechnical experience, to become an amateur geodetective, but the interesting cases are more complex – problems like locating a route for a northern railway, locating sufficient construction materials for an earth dam, planning a full-scale site investigation, evaluating the potential for groundwater seepage into an underground mine, locating a groundwater supply for a prairie townsite, evaluating the stability of a young river valley for development purposes, and many others. To attain a reasonable expertise, the photo

5

interpreter must have a framework within which to store experience – this framework is called the recognition elements.

1.3.1 Recognition elements

The major recognition elements are as follows:

(a) A landform is any surface area (large or small) which appears to be composed of a single type of earth material and which presents a particular homogeneous topographical expression. Landforms are related to their geological origin and are also called geomorphic features.

(b) A drainage pattern is a particular pattern which emerges when the natural surface drainage system (streams and gullies) are traced on a photograph. Erasable wax pencils can be used directly on photographs but it is normal procedure to use onionskin paper to trace the pattern. The basic patterns and their relation to soils and bedrock are noted on Figure 1.2. Drainage patterns are better developed above relatively impervious soils which promote surface runoff. Relatively impervious soils with good underdrainage may exhibit little or no local drainage pattern unless the topography is steeply dipping. The most common drainage pattern is dendritic and the rectangular pattern is a bedrock-modified dendritic form. Parallel drainage patterns are common on relatively flat-lying soils.

(c) Gully shapes are good indicators of soil types: steep-sided V-shaped gullies are formed in sands and gravels; short U-shaped gullies suggest silt materials; while wide smoothly rounded gullies form in clay soils. Of course, all gullies look steep on stereo pairs and until some experience is gained it is useful to plot the gully section from air photo or topographic map measurements as shown on Figure 1.3.

(d) Soil tones are expressed in relative shades on a single photo and reflect the soil moisture conditions – the darker the tone, the higher the moisture content. Coarse-grained soils generally show distinct seepage boundaries while fine-grained soils, due to high capillarity and low permeability, show irregular poorly defined boundaries or mottling. Seepage lines and soil erosion may also be observed where water seeps out of slopes. Examples are noted on Figure 1.4.

(e) Vegetation generally reflects the soil type and groundwater conditions. Pine trees thrive in well drained sand and are used to reforest blow sand areas. Tall stands of natural pines indicate sand or gravel terraces. Cedar trees, on the other hand, require abundant water. Deciduous trees generally prefer heavier soils (damp, cohesive) but the slower-growing hardwoods will thrive in lighter soils with a fairly high water table. Very wet soils generally support softwoods (poplar, ash, etc.) as well as cedars. Some stalky weeds, such as mullen, will thrive in very dry

6

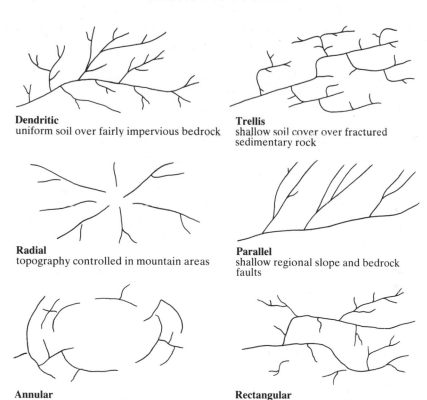

Dendritic
uniform soil over fairly impervious bedrock

Trellis
shallow soil cover over fractured sedimentary rock

Radial
topography controlled in mountain areas

Parallel
shallow regional slope and bedrock faults

Annular
controlled by a dome of fractured bedrock

Rectangular
deep soil cover on tilted rock with parallel lineations

Figure 1.2 Basic drainage patterns.

soils and scrub oak is often found when the water table is quite deep. Agricultural land use is also indicative of soil type; generally root crops are grown in sandy soils while grain crops are grown on heavier soils. The black loam of market garden areas is easily identified on photos. Examples are noted on Figure 1.4.

(f) Cultural activities often provide a basic scale for interpretation. The effects on stability and ground water of cuts, fills, dams, canals, drainage features and general land-use patterns can also be readily observed – a simple example is the drowned land that is often created by the raising of water levels in canals or storage reservoirs closely underlain by fractured rock or other continuously permeable materials.

1.3.2 Photo scales

The average scale of an air photo depends on the camera focal length and the mean flying height as shown on Figure 1.5. The average scale is $f : H$ and the

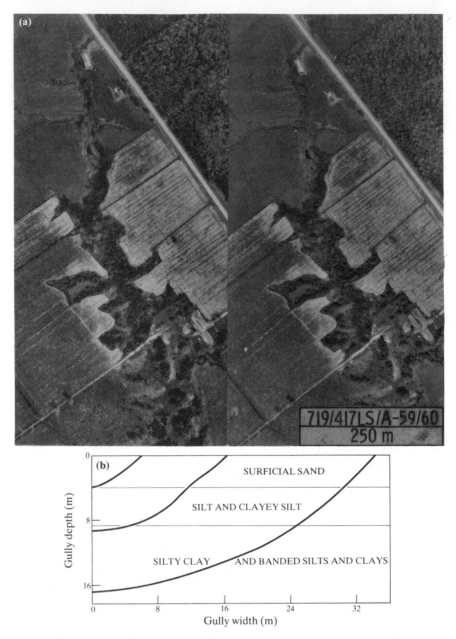

Figure 1.3 Gully shapes. (a) Gully incised through surficial sand and silt into underlying clay. (b) Typical gully sections plotted from the stereo pair in (a).(Photo courtesy of the Ministry of Transportation and Communications, Ontario.)

Figure 1.4 Contrast, vegetation and land use. (a) Silty clay ground moraine on the left has a high groundwater table indicated by the mottled tones. Deciduous trees predominate. The dark area on the right is peatlands. (b) Light tones of the well drained ancient beach sand contrasts with the darker tones of the bottom land. Contrast between photos is due to the difference in reflected light. (c) The second growth of trees in this area is mainly deciduous. Large areas have now been reforested to prevent the surficial sand from blowing. (Photos (a) and (b) courtesy of Energy, Mines and Resources, Canada. Photo (c) courtesy of the Ministry of Transportation and Communications, Ontario.)

(a)

A19411-21/22
500 m

(b)

A10907-16/17
500 m

(c)

719/417LS/A-59/60
250 m

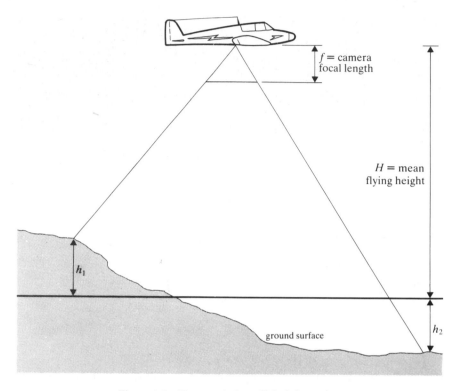

Figure 1.5   Photo scale from flight information.

scale variation across the photo will depend on the variation in topography. If $H = 2350$ m and $f = 152$ mm the average scale will be 1 : 15 460 (1 mm : 15.5 m) and if $h_1 = h_2 = 50$ m, the scale variation will be $[(2400/2300)-1]\ 100 = 4.3\%$. Time, date, camera and altimeter data generally appear along the margin of photos.

The photo scale can also be calculated from correlation with the scale of an available topographic map or from scaling some known ground dimension. A scale of 1 : 20 000 is satisfactory for most geotechnical applications – landslide studies, preliminary site (or route) selections, planning and procurement of earth materials – but a scale of about 1 : 6000 is commonly used for detailed site mapping and quantity estimates. Figure 1.6 shows the South Nation River landslide (Eden *et al.* 1971) at two different scales and the difference in detail can be noted.

It is theoretically possible to calculate vertical heights of objects from parallax measurements made directly on the photos but such measurements cannot normally be made to the required accuracy. Parallax bars and simple parallax-measuring equipment can be used without ground control to obtain vertical heights but are not recommended, due to probable operator errors,

719/417LS/2-27/28
500 m

719/417LS/B-70/71
250 m

Figure 1.6 South Nation River landslide. (Photos courtesy of the Ministry of Transportation and Communications, Ontario.)

except in the case where these are used to interpolate between known elevations. With practice it is possible to estimate slope heights to within a meter or two by comparison to known heights or to heights estimated from the shadows of tall slender objects. Slope angles should never be estimated directly (everything looks steep) but should be calculated from estimated heights and scaled horizontal dimensions.

1.4   Rock landforms

Rock landforms with a thin mantle of soils are common throughout the world: igneous rocks of Precambrian age form the basement rock over the Earth's surface and extensive outcrops such as the Canadian Shield occur on most continents; sedimentary rocks have been formed by diagenesis (diagenesis describes the development of strong cohesive bonds in mineral aggregates by cementation and precipitation to form a consolidated sediment) as a veneer over about 75% of the land mass – in North America these are mainly of Paleozoic and Mesozoic age with an abundance of weak Cretaceous formations; metamorphic rocks, a term used to describe all rocks that have been altered from their original form by granulation or recrystallization under heat and pressure, are found scattered about in the igneous formations – these are important to the mining industry, being the host rocks for most of the basic minerals. Maps of the distribution of rock types are available from government agencies. Figure 1.7 shows some major rock types. Igneous rocks often exhibit steeply dipping joint planes and metamorphic rocks are characterized by folding. Both are relatively impervious and will support surface water except where pervious fault zones extend to the surface. Sedimentary rocks generally exhibit closely spaced bedding planes and are relatively pervious. Solution channels in sedimentary limestones often result in surface collapse (sink holes).

Soils are derived from the weathering of rocks and produce various landforms depending on the method of transportation and deposition of the soil. The most abundant of the unconsolidated deposits in the northern hemisphere are the glacial landforms.

1.5   Glacial landforms

Pleistocene or glacially derived deposits cover much of the northern hemisphere and extend, in North America, across the northern United States and Canada. By grinding rocks and transporting the derived soils southward, glaciers produced much of the sorted and unsorted earth construction materials found in these areas. During glacial advance, ground moraine was laid down and drumlins (streamlined hills) were formed from the basal till material being overridden by the ice (see Fig. 1.8). During retreat, end moraines (and recessional moraines) were formed from the till while sorted outwash sands and gravels were carried away by the melt waters and deposited in glacial spillways. These granular terraces are one of the main sources of natural gravels and sands, as shown on Figure 1.9. With the retreat of the ice front a thin layer of ablation till is formed behind the end moraine or between recessional moraines. Finally, stagnant ice sheets were left to melt and produced long sinous ridges of granular material called

Figure 1.7 Rock landforms. Steeply dipping Precambrian igneous rock is separated from folded metamorphic rocks by a well drained soil landform. (Photo courtesy of Energy, Mines and Resources, Canada.)

eskers, slumped hills called kames (and kame terraces) composed mainly of sand, transverse ridges called crevasse fillings of mixed materials, and ablation moraines. Sorted materials deposited by glacial melt waters are often called fluvioglacial. Depressions from melted ice blocks, called kettle

13

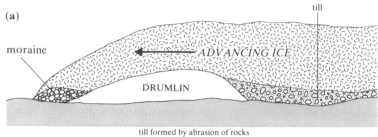

Figure 1.8 Advancing ice landforms. (a) Schematic diagram. (b) A sandy till forms the well drained drumlins and ground moraine seen on this stereo pair. (Photo courtesy of Energy, Mines and Resources, Canada.)

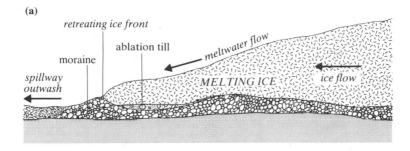

(a)

retreating ice front

ablation till

meltwater flow

moraine

MELTING ICE

ice flow

spillway
outwash

(b)

A25445-84/85

1750 m

Figure 1.9 Retreating ice landforms. (a) Schematic diagram. (b) Large gravel pits have been opened in this glacial spillway. The groundwater table is relatively high and several ponds as well as a small earth dam can be seen. End moraine and kames can be seen on the flanks of the spillway. (Photo courtesy of Energy, Mines and Resources, Canada.)

Table 1.1 Features and engineering uses of common Pleistocene landforms.

Landform	Topography	Drainage and tonal patterns	Vegetation land use	Material description	Use as borrow material
ground moraine	rolling till plains; low relief; stone piles	poorly drained dendritic; mottled dark areas; low erosion	variable; grain and pasture	variable sizes; unsorted; dense	not used extensively; good water content for easy compaction
end moraine	hummocky; irregular cross-ridges	mottled disordered; local ponds; severe erosion	woodlots; pasture; marginal farms	till with dirty sand and gravel	used for general fill purposes; generally good compaction
drumlin	aligned streamlined hills; cigar shaped	poor internal drainage; medium tone; low erosion	pasture; grain	sands to sandy tills; usually 20% clay	good borrow for dams and dikes; could require drainage
esker	long sinuous ridges of sand and gravel	well drained with darker wet flanks	wooded; pits; roadways	sand and gravel; usually clean	good source of road gravel; filter materials
kame	conical sand hills; slumped	well drained; V-gullies erodible; wind blowouts	wooded; pasture; sand pits	dry to moist sand, often includes silt fines	often dirty but sometimes a good source of sand
outwash	flat gravel terraces; kettle holes, fossil streams	high infiltration; stubby gullies; uniform with dark strings	forest; quarries	stratified clean sands and gravels	excellent granular source but often high water table
beaches	low parallel gently curved ridges	poor to good drainage; light and dark ridges	wooded; recreation	gravels and downdip sands	source of coarse sand and clean beach gravel

holes, are often found associated with these features. Some of these features are shown on Figure 1.10. Table 1.1 lists the typical composition and engineering uses of common Pleistocene landforms.

1.6 Lacustrine, marine and alluvial landforms

As the main ice sheets continued to retreat, vast quantities of melt water were produced and formed glacial lakes – sand and gravel deltas were

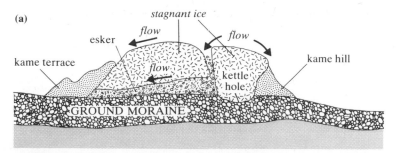

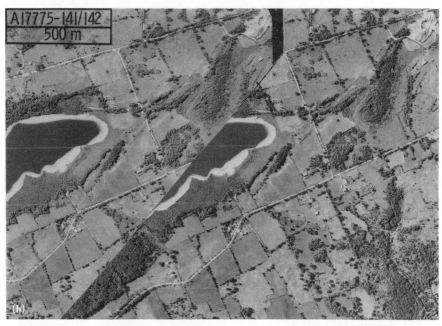

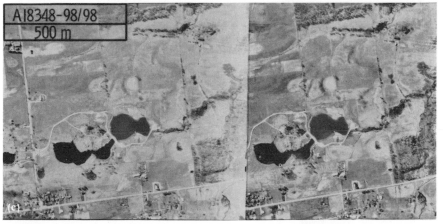

Figure 1.10 Stagnant ice landforms. (a) Schematic diagram. (b) A roadway follows the esker. Note sand pit in kame hill. (c) Kame and kettle topography. (Photos (b) and (c) courtesy of Energy, Mines and Resources, Canada.)

formed where streams deposited their sediment load into the lake and the finer materials were carried further out into the lake to form lacustrine deposits of sand beds, silt beds and varved clays (alternate layers of silt and clay that mark the seasonal effects of sediment inflow, summer settling of silts in turbulent waters and winter settling of clays in cold quiescent waters). Beaches and shorelines were formed around the glacial lakes and were left exposed when the lakes retreated to their present-day positions. Figure 1.11 shows marine sediments formed while the landmass was depressed under the mass of continental glaciers. Subsequent isostatic uplift created marine clay deposits along the northern coasts of Europe and North America. Ancient inland seas were also created as continental ice retreated from lowland areas. In this brackish environment, clay particles flocculate around silt particles and settle to produce a fairly homogeneous marine clay. Fluctuating water levels, due to ice blockages and the influx of glacial material and melt waters, also produce interbedded silts, sands and gravels in some areas. Deltaic sands, reworked by water and wind, form a thin mantle over parts of most marine and lacustrine sediments. These landforms are easily recognized because of the lack of relief, except where incised by river valleys.

Alluvial landforms are produced when river systems meander over previous deposits, altering their character and producing infilled oxbows, point bar deposits, clay plugs and back-swamp deposits as shown on Figure 1.12. Alluvium is the general term used to describe recent river valley deposits. Alluvial soils are variable and can create differential settlement and frost heave problems.

1.7 Eolian and residual landforms

Where silt floodplain deposits exist in semi-arid regions, these dry angular silt particles may be wind-transported and redeposited to form unstratified loess. Wind transport of sands from beach deposits produces dunes or dune ridges behind the beach and can produce migrating dunes if not checked by vegetation or artificial windbreaks. Sand dunes are shown on Figure 1.13.

In many parts of the world much of the native soil is residual, having been derived from in-place weathering of bedrock. Residual soils derived from Cretaceous shales often contain montmorillonite and have adverse swelling properties. Swelling minerals are also present in some of the overlying tills and result in slumping of reservoir shorelines as shown on Figure 1.13. Figure 1.13 also shows an extensive residual soil landform in northern climates – peatlands or muskeg, frozen and unfrozen. Unfrozen peat has a characteristic ribbed drainage and uniformity of surface vegetation as seen on Figure 1.4.

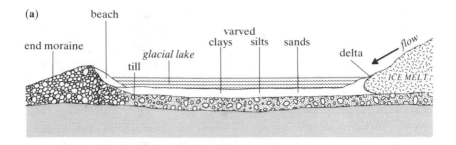

Figure 1.11 Lacustrine and marine landforms. (a) Schematic diagram (b) Ancient erosional shoreline in marine sediments. (Photos courtesy of Ontario Ministry of Natural Resources.)

1.8   Uses of air photos in earth structures engineering

Air photo studies are most useful in eliminating unfavorable dam sites or transportation routes by the recognition of hazardous terrain: rock faults, buried outwash channels or alluvial deposits, peatlands, adverse ground-

19

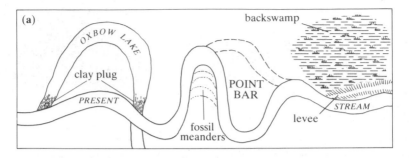

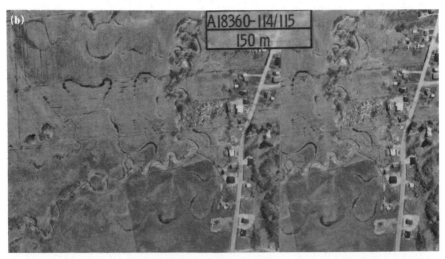

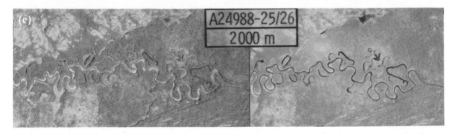

Figure 1.12 Alluvial landforms. (a) Schematic diagram. (b) Alluvial landform on lacustrine silts and clays. (c) Misfit stream reworking of granular soils. (Photos (b) and (c) courtesy of Energy, Mines and Resources, Canada.)

water conditions, landslides and subsidence. The same photo studies are ideal for preliminary site (or route) selection, aggregate resource studies, location surveys and the planning of flood control or development schemes. Some of the considerations pertinent to these studies are now outlined briefly.

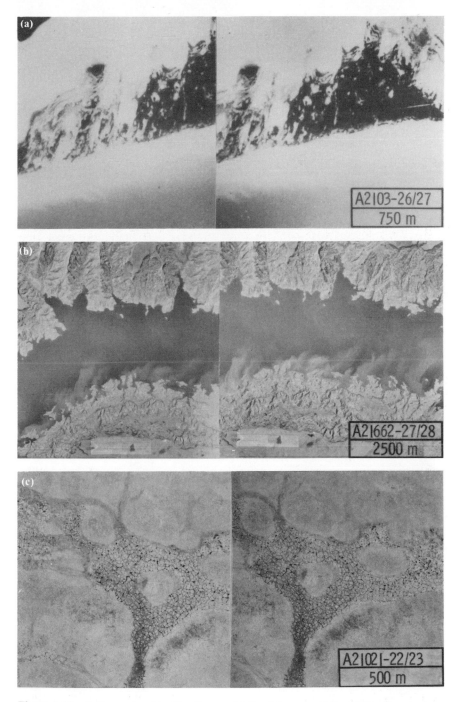

Figure 1.13 Eolian and residual landforms. (a) Sand dunes partially stabilized by pine trees. Note the shore erosion due to wave action and littoral drift. (b) Swelling residual clay. Note erosion and slumping. (c) Polygonal ice structures in frozen peat. (Photos courtesy of Energy, Mines and Resources, Canada.)

1.8.1   Slope stability studies

Stereo pairs present an overall perspective of large areas with a three-dimensional view. Boundaries of recent and prehistoric landslides can be readily detected with experience; relations between drainage and topography are more obvious than on conventional maps. Recent photographs can be compared with older ones to determine progression of critical earth movements (see, for example, Stafford & Langfelder 1971, Williams *et al.* 1979). Landsliding is a natural phenomenon along young river valleys as shown on Figure 1.14, and can occur in many forms varying from small single-block slumps, through slow progressive failures to catastrophic flows of loose earth masses or sensitive materials. The various forms of landsliding show particular characteristics that are easily recognizable, and with some experience the photo interpreter can identify subsurface materials from landslide characteristics. It is often possible to survey river valleys using air photos and profiling instruments or topographic maps as part of a terrain

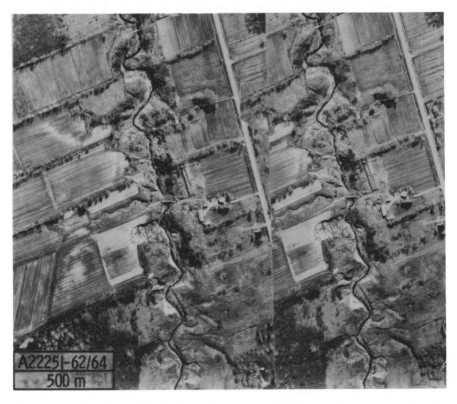

A22251-62/64
500 m

Figure 1.14 Erosion and landsliding. Toe erosion and high ground water promote valley widening by back-and-forth landslide activity. (Photo courtesy of Energy, Mines and Resources, Canada.)

evaluation study – from these surveys a histogram of the number of stable and unstable slopes can be plotted against slope angle to correlate with detailed slope stability analyses.

Air photos are also useful in investigating the causes of major landslides, in establishing mechanisms for analysis and in identifying similar slope stability problems before they develop. High-level photographs can be used to locate active landslides in mountain areas as exemplified on Figure 1.15. Greater detail can then be obtained from low-level photos as noted with regard to the major earthflow shown on Figure 1.6. Two common landslide types are shown on Figure 1.16, and Figure 1.17 shows a large rockslide which took place on steeply dipping bedding planes in weak sedimentary rocks.

Figure 1.15 Debris flow or mudflow. Active downslope movements at Drynoch landslide create maintenance problems for transportation routes. (Photo courtesy of Energy, Mines and Resources, Canada.).

23

Figure 1.16 Common landslide types. (a) Block slump in a lacustrine sediment. (b) Rotational landslide. Cedar trees indicate abundant water. (Photos courtesy of Energy, Mines and Resources, Canada.)

1.8.2 Ground subsidence

The most common natural subsidence is the sink hole which occurs when surficial unconsolidated material subsides into solution cavities and caves in soluble rocks. Classical examples are the subsidence in the karst topography of Yugoslavia and South Africa. In North America, sink holes in limestones vary from shallow depressions to pits as deep as 40 m. Recent caving of surficial sands into an underlying limestone cavern in Florida produced a crater 200 m in diameter by 50 m deep.

Ground subsidence due to extraction of ground waters has long been

Figure 1.17 Rockslide at Frank, Alberta, Canada. Large translatory landslide moved 10^8 tonnes of rock. (Photo courtesy of Energy, Mines and Resources, Canada.)

known. Portions of the city of Houston, Texas, have subsided several meters due to the extraction of water from underground sources. Subsidence in the Long Beach, California, oilfield has influenced an area of about 50 km^2 with maximum settlement of about 8 m. Subsidence rates are being reduced by pumping salt water into the oil-bearing formation under pressures ranging up to 8000 kPa. Over 100 million dollars have been spent on rehabilitation work to engineering structures and services.

In mining operations, the practice of backfilling underground openings (stopes) helps to keep subsidence to a minimum. Where underground workings are left unsupported, subsidence is inevitable. Figure 1.18 shows an example of subsidence, and subsidence analysis is discussed in Chapter 6.

1.8.3 Dam site selection

Landforms are referred to as being consolidated, partially consolidated or unconsolidated. Softer and less resistant rocks are worn away by weathering and erosion to leave the more resistant, better consolidated materials as

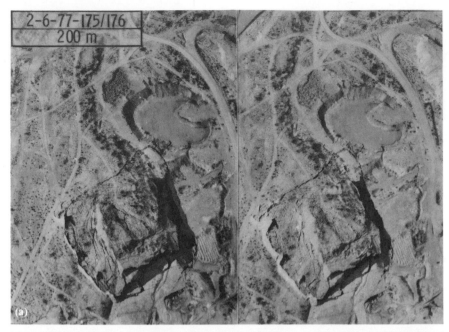

Figure 1.18 Ground subsidence. (a) Subsidence due to caving into an underground mine. (b) Surface scarps due to caving in an underground mine. (Photo (a) courtesy of US Department of the Interior, Bureau of Mines. Photo (b) courtesy of John D. Smith Engineering Associates Ltd.)

positive landforms. Geological structures of importance to the engineer such as jointing, fracturing, folding and faulting are better exposed in consolidated materials. These materials will often form the foundation for an earth dam and these features, which will affect the watertightness of the foundation, can often be detected on photos. Where surface drainage is bedrock-controlled, fault locations may often be inferred by the location or dislocations of rivers and streams as shown on Figure 1.19. The area shown on Figure 1.19 is not an area of seismic activity, and a hydroelectric power facility was constructed as shown on Figure 1.20.

Unconsolidated materials are usually sediments deposited by streams, glaciers, wind and sometimes by volcanic activity. The engineer will search for a source of borrow material in such unconsolidated materials. He must therefore make every effort to determine how the material was deposited and its likely composition. Initial quantity estimates can also be made from air photos.

In dam site investigations, ground water and groundwater movements are also of primary importance. Detailed air photo studies of specific formations (material identification and drainage patterns) some distance from the actual dam site may be necessary to supplement hydrological studies (usually carried out using large-scale topographic maps and well records). Changes in the drainage patterns resulting from reservoir filling may have detrimental effects on the stability or cultural use of adjacent land areas. Groundwater flow concentrations that might produce artesian pressures or internal erosion are of particular concern. Figure 1.21 shows two examples of adverse groundwater pressures.

Reconnaisance surveys are always necessary to confirm photo interpretations but the photos provide the basic information for planning preliminary site investigations. Dam site selection and site investigation are discussed further in Chapter 5.

1.9   Other remote sensing and probing techniques

Infrared scanning and radar imagery have been mentioned earlier – both are used in airborne terrain scanning to produce mosaic prints. Infrared reflects the temperature (heat loss) from objects and is used to detect plant disease, thermal pollution in lakes, groundwater pollution from dumps, objects hidden beneath vegetation, heat loss from buildings and anything else that is slightly hotter than its neighbors. Very small temperature differences can be detected, but at normal thermal settings this technique can produce photo prints that are very similar to ordinary air photos. The same recognition elements are also used in interpreting radar imagery, a technique which is used mainly in bedrock areas where weather conditions make ordinary air photos difficult to obtain. Bottom contours of offshore areas are also

27

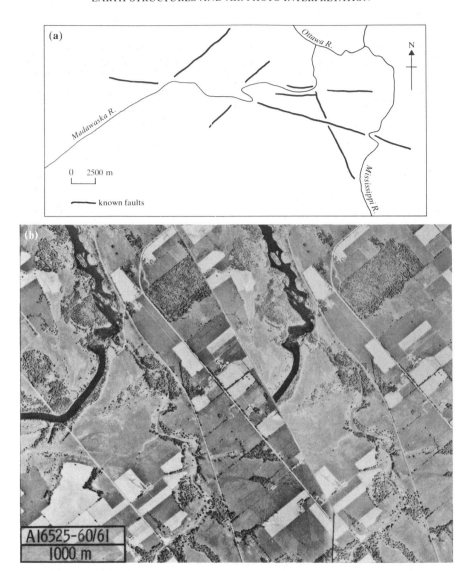

Figure 1.19 Bedrock and drainage. (a) Schematic diagram. (b) Land drainage is controlled by both surface and bedrock topography. Rivers and streams are often located above buried bedrock valleys or rock faults. Note location of small stream and the rock outcrops in the river. (Photo (b) courtesy of Energy, Mines and Resources, Canada.)

produced remotely by ships using radar and sonar scanning equipment.

Geophysical methods for determining the depth to bedrock and the stratigraphy of overburden have been in use since the early part of this century – electrical resistivity and seismic refraction methods have been successful in mapping shallow geological profiles to obtain the depth to

Figure 1.20 Arnprior Dam, Ontario, Canada. (a) The concrete spillway and powerhouse are constructed on rock foundations. The right abutment is a retaining wall transition to the earth embankment but the embankment forms a wrap-around transition to the concrete on the left. (b) A saddle dam prevents inundation of the old flow channel or fault-related depression (saddle). (Photos courtesy of Energy, Mines and Resources, Canada.)

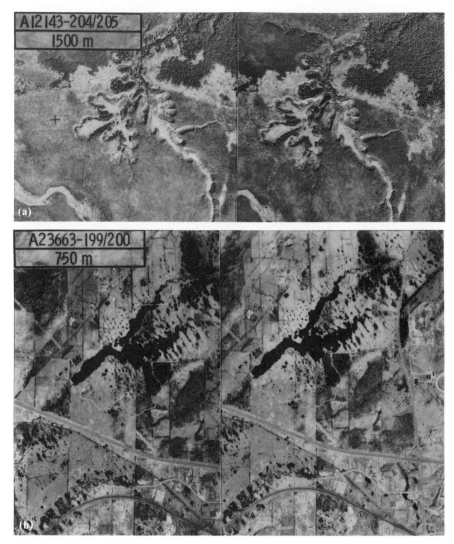

Figure 1.21 Sapping and artesian groundwater pressures. (a) Internal erosion (sapping) in sand due to horizontal flow concentrations from free water at uplands elevation. (b) Small earth dam with sheet pile and rockfill overflow section located in glacial channel underlain by pervious lacustrine sands. Groundwater flow concentrations due to adjacent lakes connecting through pervious limestone rocks create artesian pressures in the channel materials. (Photos courtesy of Energy, Mines and Resources, Canada.)

bedrock and indications of any buried channels, faulting and rock weathering, position of the water table and some estimates of physical material characteristics. In some cases, bodies of discontinuous permafrost have been distinguished from bedrock and surrounding unfrozen materials. For

30

details of these methods the reader is referred to texts on geophysical methods. Some recent advances in geophysical methods (including acoustic and impulse radar) are described in ASCE (1974a).

Ground probing is very effective in soft soils, such as lacustrine deposits or peatlands; a small (25 mm internal diameter) open sampler (called a peat sampler) is attached to a light pipe and pushed into the ground, and a sample for identification purposes is recovered at any desired depth by pulling up on

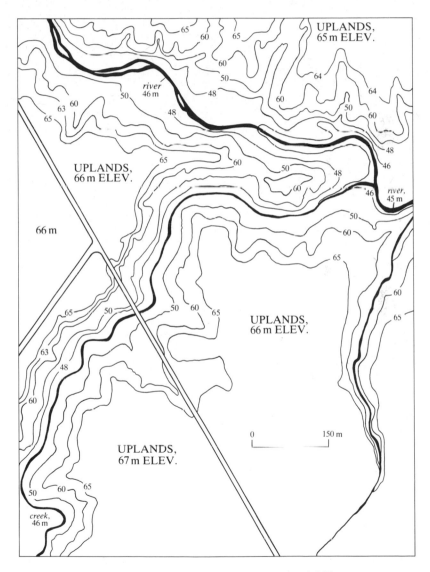

Figure 1.22 Contours of Green Creek area (1967).

31

A23217-80/83
200 m

Figure 1.23 Stereo pair of Green Creek area (1974). (Photo courtesy of Energy, Mines and Resources, Canada.)

the pipe to open the sampler, pushing to fill the sampler and extracting the string of pipe and sampler. A two-man crew can complete 10 to 20 m per hour of depth probing in soft soils (with sampling every 2 m). An auger attachment (ordinary wood auger bit welded to a pipe section) is necessary for penetrating weathered surface crusts or thicker layers of sand or silt and for confirming when the sounding has reached bedrock. In stiffer soils, ground probing involves mechanical driving of a cone or standard penetration sampler. Test details and interpretation of the standard penetration test are

32

described in most soil mechanics texts. Penetration tests can be carried out at rates between about 10 and 50 m per hour and thin lenses of materials such as sand, gravel or even ice, which may be very important to groundwater movement, can be detected. For route and site selection, remote sensing and ground probing can provide a valuable and efficient link between available published information and detailed site investigation.

1.10 Problems on air photo interpretation

Problem 1   (a) Figures 1.22 and 1.23 show a section of Green Creek, near Ottawa. Prepare a histogram of slope stability versus slope angle by plotting the average slope angle for each of the five or six landslides evident on the photo as slope failures (the slope angle prior to failure must be plotted) and average slope angles for several of the steeper stable slopes. What is the critical slope angle for slopes of 20 m height in this material under the prevailing conditions?

              (b) Describe the landform and groundwater conditions in this area, listing all confirming recognition elements.

              (c) Comment on the potential risk of slope failure where the farm buildings exist between the bridge structure and the roadway intersection. What minimum setback from the top of the bank would you suggest as appropriate for new buildings in this area?

A22223-29/30

500 m

Figure 1.24 Slopes and residential development. (Photo courtesy of Energy, Mines and Resources, Canada.)

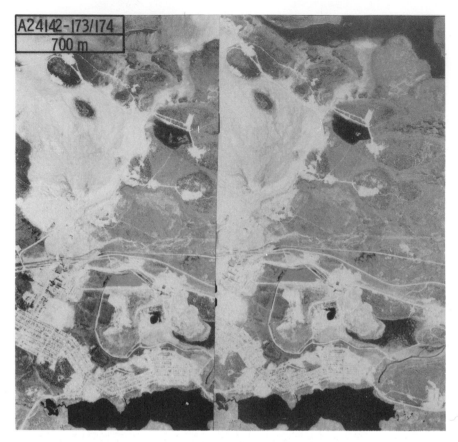

A24142-173/174
700 m

Figure 1.25 Mine tailings and waste dumps. (Photo courtesy of Energy, Mines and Resources, Canada.)

Problem 2 Figure 1.24 shows a typical residential development on a ravine. Estimate the slope height and average slope angle behind these houses. Identify the material type and comment on the groundwater conditions. Are any slope stability problems evident? If so, describe these.

Problem 3 Figure 1.25 shows an open pit mine and associated facilities. Waste rock is being dumped to the right of the open pit and a tailings disposal area can be seen in the upper portion of the stereo pair. What is the purpose of the series of dikes surrounding the open pit? What type of rock outcrops in this area? What is the soil type? Estimate the maximum height and slope angle of the dike at the tailings discharge point. What type of earth material was used to construct this dike?

Figure 1.26 Landslide identification. (Photo courtesy of Energy, Mines and Resources, Canada.)

Problem 4 Figure 1.26 shows an area where a number of landslides have taken place. Identify the landforms, soil type and vegetation on this stereo pair. Comment on the groundwater levels in this area. Identify three different forms of landsliding and comment on which form is ancient and which form is recent in this area.

419/417LS/4-50
300 m

Figure 1.27 Drainage pattern. (Photo courtesy of the Ministry of Transportation and Communications, Ontario.)

Problem 5   Identify the basic drainage pattern shown on Figure 1.27 and comment on the character of the underlying soils and bedrock.

Problem 6   (a) Describe the landforms shown on the upper third, middle portion and lower third of Figure 1.28.
(b) Briefly describe the origin of these landforms.
(c) Suggest what type of material would be found in each landform identified and comment on the use of these materials in the construction of a small earth dam.

Problem 7   (a) Identify the landform and type of soil in the central portion of Figure 1.7.
(b) Comment on the past and present stability of areas adjacent to the Drynoch landslide on Figure 1.15.
(c) List the recognition elements which indicate that the material shown on Figure 1.21 is sand. In what type of landform is the sand contained?

Problem 8   (a) What is the porous-looking material in the reservoir shown on Figure 1.20?

A25445-106/107
1750 m

Figure 1.28 Borrow materials. (Photo courtesy of Energy, Mines and Resources, Canada.)

(b) What process caused the depression seen on the prominent point of land jutting into the reservoir shown on Figure 1.20?

(c) Comment on the erodibility of the reservoir slopes shown on Figure 1.20 compared to those shown on Figure 1.13.

(d) Estimate the height and the downstream slope angle of the saddle dam shown on Figure 1.20.

(e) What local environmental problem might be created by the reservoir shown on Figure 1.20?

2 Earth mechanics in earth structures engineering

After completion of preliminary studies using available information sources and remote sensing or probing techniques, a site investigation is carried out to obtain detailed stratigraphic information and samples for identification and testing purposes. The two most important considerations in any site investigation are as follows:

(a) Not to miss important detail: many examples of unanticipated construction problems and failures of earth structures have resulted because some important detail, such as thin weak layers, pervious layers, artesian pressures, fracturing or other discontinuities, or lateral variations in stratigraphy or lithology, has been missed during site investigations.
(b) To obtain the correct material properties: for laboratory testing, representative undisturbed samples must be obtained; alternatively, or as an additional source of data, representative field tests must be conducted.

Even continuous sampling can miss thin seams at some depths in the borehole because some core loss or disturbance accompanies all sampling methods. The question of borehole spacings (how many boreholes are required) is always debatable but preliminary studies should be used to guide this decision. One thing is certain – the cost of an efficient and reliable site investigation is not always proportional to the cost of the structure to be built but depends, as well, on the complexities of the terrain over which the structure will be built. Usually there are sufficient funds available for a detailed site investigation for a large project but many smaller projects have encountered difficulties, failed or required costly remedial measures due to the lack of a detailed investigation. As a site investigation progresses, the findings should be compared with expectations based on the preliminary studies and the program should be altered, if necessary, on the basis of this comparison.

The earth structures engineer should be familiar with modern site investigation and testing techniques including boring and sampling methods (summarized on Table 2.1), classification (briefly summarized on Table 2.2), field tests (briefly summarized on Table 2.3), laboratory testing (briefly summarized on Table 2.4) and basic earth mechanics concepts. This chapter is intended only to summarize some of the more important concepts relevant

Table 2.1 Brief summary of drilling and sampling methods.

Type of earth material	Applicable drilling methods	Applicable sampling methods
rock	diamond drill; percussion drills; rotary drills; jet pierce drill	core barrels used with diamond drilling
coarse-grained soils	shallow augering or wash boring with driven casing; hollow stem auger drill; drilling fluid may be required	drive sampler with core catcher; split spoon; augers; Denison sampler
mixed soils (tills) with stones	shallow augering; hollow stem augering; diamond drill	pitcher sampler; block samples from excavations
mixed soils (tills) without stones	wash boring with casing; hollow stem augering; continuous sampling	piston samplers; sharp open Shelby tubes (driven); block samples from excavations
overconsolidated clays	continuous sampling; wash boring; hollow stem augering	piston samplers; sharp open Shelby tubes (pushed); block samples from excavations
insensitive normally consolidated clays	wash boring with or without casing; continuous sampling displacement by piston sampler	piston sampler; sharp open Shelby tubes (pushed); block samples from excavations
sensitive clay and silt sediments	continuous sampling; wash boring with casing	Osterberg sampler; Geonor sampler; Swedish foil sampler; special block sampling

to earth structures engineering and to provide some data on typical properties of earth materials.

2.1   Strength and deformation of earth materials

The capability of unreinforced earth masses to resist tensile stresses is minimal, although reinforcing by rock bolts or other metal reinforcement, by cementation or by maintaining pore-water pressures less than the surrounding ambient pressure can provide an artificial permanent or temporary resistance to tension. Compressive stresses and strains are, therefore, defined as positive and the strength of an earth material is generally defined in terms of the maximum stress difference, $\sigma_1 - \sigma_3$ (often termed the deviatoric stress), either at failure or at some limiting strain. The student of mechanics will note that there are three principal stresses ($\sigma_1, \sigma_2, \sigma_3$) and strains ($\epsilon_1, \epsilon_2, \epsilon_3$) in a stressed material and that the above definition of strength ignores any possible effect of the intermediate principal stress on shearing resistance. It

Table 2.2 Engineering classification of earth materials.

(a) Unified classification system for soils

		Group symbol	Fines <0.075 mm	$\dfrac{D_{60}}{D_{10}}$	PI of fines
Coarse grained has more than 50% larger than 0.075 mm. Gravels have largest proportion >2 mm. Sands have largest proportion between 0.06 mm and 2 mm	well graded sandy gravels with little or no fines	GW	0–5	>4	
	poorly graded gravels and sandy gravels	GP	0–5	<4	
	silty gravels and silty sandy gravels	GM	>12		<4
	clayey gravels and clayey sandy gravels	GC	>12		>7
	well graded sand with gravel sizes but no fines	SW	0–5	>6	
	poorly graded sands with little fines	SP	0–5	<6	
	silty sands with little true clay minerals	SM	>12		<4
	clayey sands with plastic fines	SC	>12		>7
Fine grained has more than 50% smaller than 0.075 mm. Silts and clays are classified using the plasticity chart shown on right	inorganic silts and silty sand mixtures	ML			
	inorganic clays and silty or sandy clays	CL			
	organic silts and organic clays	OL			
	inorganic silts of high plasticity	MH			
	inorganic clays of high plasticity	CH			
	organic clays of high plasticity	OH			

Plasticity chart (axes): PI (vertical, 0–60) versus w_L (horizontal, 0–80). Regions labelled: CL, CH, CL-ML, MH OH, ML OL, A line; 7, 4 marked on PI axis.

Dual symbols are used as required, e.g. SW–SM with fines 5–12%.
Highly organic soils such as peats and muskegs are given symbol Pt.

(b) Classification of rocks on the basis of strength and discontinuities

Strength range (unconfined) (MPA)		Rock quality RQD		Spacing of fracture joints and planes		Rock mass classification
very high	>200	very good	>90	very wide	>3 m	intact, solid
high	100–200	good	75–90	wide	1–3 m	massive
medium	50–100	fair	50–75	moderate	0.3–1 m	blocky, seamy
low	25–50	poor	25–50	close	0.05–0.3 m	fractured
very low	<25	very poor	<25	very close	<0.05 m	shattered

Rock quality designation

$$RQD = \frac{\Sigma \text{ intact core pieces} > 100 \text{ mm in length}}{\text{length of hole drilled}} \times 100\%$$

was introduced by Deere (1964) to relate rock quality obtained by double-barrel diamond core drilling to rock performance. For further details see Barton *et al.* (1974).

Table 2.3 Brief summary of common field tests.

Field test and brief description of test	Relative cost and difficulty of test	Types of soils in which test is useful	Information derived from test
Field vane: A thin-bladed vane is rotated in the soil. The vane can be advanced with or without a borehole, depending on stiffness of soil and stratigraphy	Simple test. Cost depends on need for borehole to advance through non-cohesive layers	Used only in cohesive (clayey) soils which are at or near full saturation. Remolded strength also measured	The undrained strength (Cu) is derived from the torque required to rotate the vane. Soil sensitivity obtained
Standard penetration test: A standard 51 mm diam. by 813 mm long sampler is driven by a 63.5 kg mass free-falling 0.76 mm. Blow count for the second 152 mm of penetration is recorded	Common test and is relatively inexpensive. Disturbed samples are obtained	Used mainly for testing cohesionless soils that are difficult to sample. Relative strength of cohesive soils can also be obtained	Empirical correlations with density, frictional strength and bearing capacity used for design in cohesionless soils
Cone penetration tests: A pointed cone is pushed (static test) or driven (dynamic test) into the soil. End resistance is measured or blows counted	Simple and inexpensive test. Some static cones have friction sleeves and pore-water pressure measurement	Static test generally used in cohesive soils or tills. Static test with friction sleeve used in layered soil systems	Stratigraphy (layering), relative stiffness of clay and approximate undrained strength. Empirically correlated with friction and density in sands
Pressure meter: A cylindrical diaphragm is expanded into the walls of a borehole. The volume versus pressure relation is measured	Relatively complex and expensive. Subject to testing errors if used by inexperienced operator	Can theoretically be used in all soils but soil homogenity is desirable	*In situ* horizontal stress (approximate) and stiffness modulus. Strength calculations based on analytical assumptions
Borehole shear test: This uses friction sleeves or various other devices to shear soil along the boundary of the borehole	Intermediate to relatively simple test. Cost mainly in borehole preparation	Not commonly used in soil mechanics but of interest in silts and sands where sampling is difficult	The strength parameters c and ϕ can be derived from a number of tests (assumes isotropy)

is often assumed that earth materials are isotropic (laboratory strength tests being carried out on samples obtained from vertically oriented boreholes) and, in most earth materials, this assumption is of greater significance than ignoring the effects of σ_2. These effects will be discussed later in this chapter and will be further developed in subsequent chapters.

41

Table 2.4 Brief summary of common laboratory tests.

Laboratory test and brief description	Type of material on which test is most commonly used	Information obtained from test
Sieve analyses: Material is separated by mechanical shaking or washing through a series of sieve opening sizes	All soils. Dry sieving for coarse-grained soils and wet sieving for fine-grained soils (followed by hydrometer test on < 0.075 mm sizes)	Dry mass retained on each sieve size used to obtain grain size distribution
Atterberg limit tests: A drop cup is used to obtain the liquid limit and the plastic limit is obtained by rolling	Fine-grained soils or separated fines. The fall cone test may be used as an alternative for obtaining the remolded limits	Remolded characteristics for classification and empirical correlations. Range of plastic water contents obtained
Compaction test: A mold and and drop hammer (standard or modified) are used to compact samples according to test specifications. Dynamic densification	All compactable materials less than 15 mm in size. Stone correction can be done if small proportion above 15 mm. Used to specify field conditions for compaction	Relationship between dry density and water content for compaction energy used. Field density tests used to monitor field compaction
Compression test: Material contained in a short cylindrical metal ring is compressed under static stresses Consolidation test: As above, with substantial time delays between stress increases to allow pore-water pressure dissipation	Natural undisturbed and compacted materials. Sample depth and diameter should be large with respect to maximum particle size. Natural undisturbed and saturated compacted samples. Pore-water pressure measurements remote from drainage boundary optional	Void ratio versus axial stress plot to obtain coefficient of volume change, preconsolidation pressure and compression index. Compression versus time (root time or log time) to obtain time to full consolidation and coefficient of one-dimensional consolidation
Permeability tests: Water is allowed to flow through sample under constant head or falling head. Hydraulic gradient controlled	All pervious materials. Apparatuses variable depending on sample. Special end fittings are available for compaction molds. Triaxial cell used for undisturbed samples	Coefficient of permeability, also called coefficient of hydraulic conductivity (and variations with changes in void ratio)
Direct shear (shear box) tests: A thin shear plane is formed by relative displacement of halves of the box under drained conditions	Compacted samples and granular materials. Also oriented joints in rock cores. Multiple reversals of displacement used to develop a residual shear plane in clays	Shear stress versus effective normal stress plotted to obtain c', ϕ' and residual ϕ' (after reversals). Force–displacement relation
Triaxial tests: Cylindrical sample in impervious membrane is subjected to drained or undrained stress changes. Boundary displacements and pore-water pressure measured	Undisturbed cohesive samples and prepared samples (compacted or loose). Cyclic or repeated loads can be applied. Wide range of compression and extension stress paths	Stress–strain relations and strength envelope (Cu, c', ϕ'). Pore-water pressure parameters; static and cyclic behavior
Plane strain and simple shear tests: Specialized strength and stress–strain tests used mainly in research	Undisturbed or laboratory prepared samples. Cyclic and repeated loads can be applied	Soil behavior under plane strain (no lateral strain) and under simple shear (distortion with rotation of principal stresses)

2.1.1 Effective stresses in earth materials

The single most important concept in understanding the strength and behavior of earth materials is provided by defining the effective stresses within the solids fabric as

$$\sigma' = \sigma - u \tag{2.1}$$

where σ is the total stress and u is the total pore-water pressure. When, for reasons of artesian groundwater flow or rapid compressive volume change tendencies, the excess pore-water pressure increases above the hydrostatic value of $u = h_w \gamma_w$, the effective stress decreases proportionally. Since the shearing resistance of a granular material such as sand depends mainly on the friction derived from intergranular (effective) stresses, the material is weaker when the pore-water pressure is high and stronger when the pore-water pressure is low. Quicksand results when $u \to \sigma$ and sand can be stabilized by suction wells where $u < h_w \gamma_w$. Partly saturated materials have capillary water under suction and these soils can be weakened by increases in water content. The settlement of structures is also related to changes in effective stresses as applied loads are transferred throughout the material fabric.

This principle of effective stress was developed by Karl Terzaghi half a century ago (see Terzaghi 1943) and has allowed geotechnical engineers to account for the effects of water pressures generated in earthworks by construction activities and by impoundment dams. In earth materials the horizontal effective stress may be greater than or less than the vertical effective stress depending on the geological history of the materials. A coefficient of earth pressure at rest is used to relate the principal effective stresses beneath a horizontal surface in the form $\sigma'_h = K_0 \sigma'_v$. Recent application of the effective stress principle to water pressures generated between automobile tires and impervious roadway surfaces has contributed to the development of the particulate tread of rain tires.

2.1.2 Mohr–Coulomb strength criterion

The typical data on Figure 2.1a can be well approximated by the formula

$$\tau_f = c' + \sigma'_N \tan \phi' \tag{2.2}$$

The typical data on Figure 2.1b can be well approximated by the equation

$$\left(\frac{\sigma_1 - \sigma_3}{2}\right)_f = d' + \left(\frac{\sigma'_1 + \sigma'_3}{2}\right)_f \tan \psi'$$

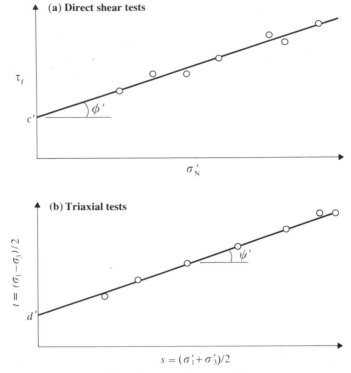

Figure 2.1 Data interpretation.

These data can be directly related if it is assumed that the Mohr–Coulomb failure criterion applies to the material. This criterion states that the material will fail on the plane of maximum obliquity and is similar to solid block frictional sliding, as shown on Figure 2.2. For this reason, the angle ϕ' is known as the angle of internal friction, or simply friction angle, of the soil.

The plane of maximum obliquity is oriented at $(45 + \phi'/2)$ degrees to the plane of the major principal stress as shown on Figure 2.2, and soils are generally observed to develop shear planes close to this angle in the triaxial test. From Figure 2.2 it can be shown that

$$d' = c' \cos \phi'$$

$$\tan \psi' = \sin \phi' \tag{2.3}$$

Then the triaxial data are expressed mathematically as

$$\left(\frac{\sigma_1 - \sigma_3}{2}\right)_f = c' \cos \phi' + \left(\frac{\sigma_1' + \sigma_3'}{2}\right)_f \sin \phi' \tag{2.4}$$

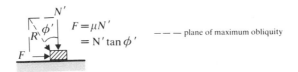

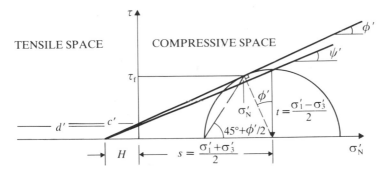

Figure 2.2 Mohr–Coulomb criterion.

Thus c' (called the apparent cohesion) and ϕ' can be readily calculated from the best-fit line through the test points as plotted on Figure 2.1b. This method of plotting simplifies data analysis since it is not necessary to plot the Mohr stress circles and it allows linear regression to find the best-fit Mohr–Coulomb criterion. The maximum shear stress, $t = (\sigma_1 - \sigma_3)/2$, is independent of the pore-water pressure and the effective average stress, $s = (\sigma'_1 + \sigma'_3)/2$, is equal to $(\sigma_1 - \sigma_3)/2$ plus the all-round pressure (σ_3) minus the pore-water pressure. Values of (t, s) represent the top point of the Mohr stress circles and trace an effective stress path for an element of material subjected to axially symmetric or plane strain deformation.

It is important to appreciate that the Mohr–Coulomb equation is a theoretical strength criterion and is a straight line, while test results form a strength envelope which is not necessarily a straight line and must never be extrapolated as such. Indeed, extrapolation to the tensile stress space on Figure 2.2 would indicate that any material exhibiting apparent cohesion (c') would have a significant tensile strength $(\sigma_t = H)$ and this is known to be erroneous. The Griffith (1924) theory of crack propagation is used in rock mechanics to explain how tensile stress propagations can develop at low average stresses and produce a curvature of the strength envelope. In dense granular or fractured earth materials the experimental strength envelope is observed to be curved, as shown on Figure 2.3 (Ladanyi & Archambault 1970, Eden & Mitchell 1970). Even in loose sands (which exhibit $c' = 0$) the value of ϕ' may be reduced at high average stresses due to grain crushing (Bishop *et al.* 1965). Laboratory testing must be conducted over the appropriate range of average effective stress (or normal effective stress) in order to

45

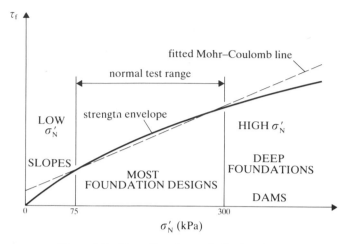

Figure 2.3 Strength and normal effective stresses.

establish the correct strength parameters, c' and ϕ', for a given design problem.

Unconfined tests ($\sigma_3 = 0$) are often conducted on rock cores and can measure either the unconfined compressive strength (if the sample crushes) or the shear strength on a particular plane if the sample fails by shear on a weak plane. In the latter case, a Mohr circle construction is used to find τ_f and σ_N' on the observed plane of failure. Unconfined quick tests can be used to give a quantitative evaluation of statistical or directional variations in the undrained strength of fine-grained (cohesive) soils, but the unconfined strength may be less than the theoretical value of $[\sigma_1]_f = 2Cu$, due to the effects of fissuring or to tensile stresses induced by end restraint at the loading platens.

2.1.3   Basic observations from undrained soil behavior

By analyzing results of undrained triaxial tests on saturated soils, Skempton (1954) discovered that the pore-water pressure increase could be related to total stress increases as

$$\Delta u = B[\Delta\sigma_3 + A(\Delta\sigma_1 - \Delta\sigma_3)] \tag{2.5}$$

where A and B are experimental constants. For three-dimensional stress changes, this equation can be expanded to

$$\Delta u = B\{ \tfrac{1}{3}(\Delta\sigma_1 + \Delta\sigma_2 + \Delta\sigma_3) + A'[(\Delta\sigma_1 - \Delta\sigma_2)^2 \\ + (\Delta\sigma_2 - \Delta\sigma_3)^2 + (\Delta\sigma_3 - \Delta\sigma_1)^2]^{1/2}\}$$

where A' will be a constant different from A.

46

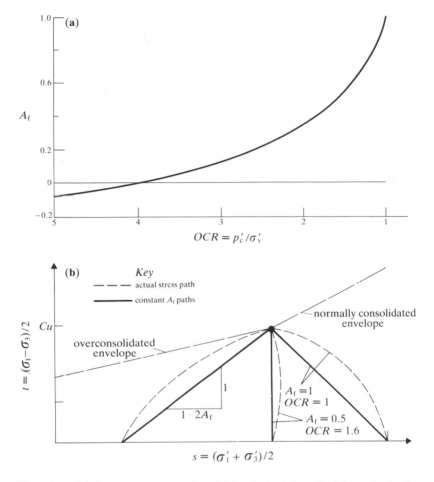

Figure 2.4 (a) Pore-water pressure A_f and (b) undrained strength of fine-grained soils.

The value of B is theoretically equal to unity for a saturated soil, but has been found to be slightly less than unity in some dense soils, in which small quantities of entrained gas can increase the compressibility of the water to approach that of the soil structure. The value of A varies from close to the theoretical elastic value of $A=1/3$ near the start of a test to a particular value, A_f, at failure which depends mainly on the density of the soil being tested. A typical relationship between A_f and the overconsolidation ratio (OCR) for a clay soil is shown on Figure 2.4a. The basic purpose of the Skempton pore-water pressure equation is to provide a mathematical basis for predicting the generation of pore-water pressure in a soil during construction activities, but these observations can be used to develop other general relations which show the following:

47

(a) The effective stress path is unique for a given saturated soil when sheared undrained from a given initial state. This means that the undrained strength of that element of soil is theoretically a unique value (Cu) and explains why field tests (such as the vane test) can give the same value of strength as triaxial tests even though the loading paths are much different.

(b) The undrained strength of a saturated fine-grained soil depends mainly on its void ratio. The A_f value changes with OCR such that all effective stress paths terminate near the same point on the failure envelope, as shown on Figure 2.4b (minor differences being due to the minor changes in void ratio as the soil swells to the different initial overconsolidated states). A major decrease in void ratio due to applied stresses exceeding the preconsolidation pressure, p'_c, will decrease the void ratio and increase the undrained strength of the soil. Ample experimental proof of this behavior is given by Henkel (1960) and Roscoe *et al.* (1958).

Since the undrained strength is given with reference to Figure 2.5. as

$$Cu = t_0 + \Delta t = t_0 + \Delta s/(1 - 2A_f)$$
$$= c' \cos \phi' + s_f \sin \phi'$$

and

$$t_0 = \sigma'_v (1 - K_0)/2 \qquad s_0 = \sigma'_v (1 + K_0)/2$$

then

$$\frac{Cu}{\sigma'_v} = \frac{c' \cos \phi'}{\sigma'_v [1 + (2A_f - 1) \sin \phi']} + \frac{\sin \phi' [K_0 + A_f(1 - K_0)]}{1 + (2A_f - 1) \sin \phi'} \qquad (2.6)$$

For normally consolidated clay soils, $c' \simeq 0$, $K_0 \simeq 1 - \sin \phi'$ and

$$\frac{Cu}{\sigma'_v} \simeq \frac{\sin \phi' [1 + \sin \phi' (A_f - 1)]}{1 + (2A_f - 1) \sin \phi'} \qquad (2.7)$$

If A_f is approximated by unity for a normally consolidated soil, then

$$\frac{Cu}{\sigma'_v} \simeq \frac{\sin \phi'}{1 + \sin \phi'} \qquad (2.8)$$

The minimum undrained strength ratio is approximated for an unstructured normally consolidated soil (Skempton 1957) as

$$\frac{Cu}{\sigma'_v} = 0.11 + 0.37PI \qquad (2.9)$$

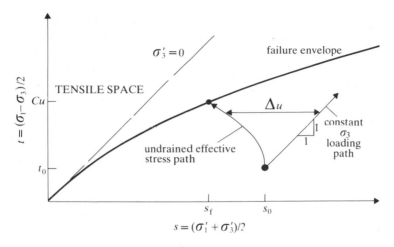

Figure 2.5 Effective stress and undrained strength.

2.1.4 Typical behavior of earth materials

Loose granular soils exhibit positive volume changes (decrease in volume due to shear, producing a plastic type of stress–strain curve and a linear Mohr–Coulomb envelope with $c' = 0$. Dense granular materials dilate (increase in volume) during shear, producing a peak strength at maximum dilation rate and a post-peak reduction in shearing resistance as a critical void ratio is attained in the zone of shear. The Mohr–Coulomb failure envelope for a dense granular or highly fractured material is generally found to be gently curved, as shown on Figure 2.6.

A loose sand can distort continuously and absorb much greater energy per unit volume (approximated by the area under the stress–strain curve) than a dense sand without exhibiting a rupture surface. This property of loose sands can be useful in particular design situations and allows design confidence at lower factors of safety. Loose sands are, however, subject to large settlements under load and are more susceptible than dense sand to liquefaction failures.

Normally consolidated clay soils exhibit a plastic behavior, with strength being mainly dependent on voids ratio (or water content since these soils are saturated to near surface in nature). When drainage develops during shear the void ratio will decrease with shearing stress increase and a normally consolidated failure envelope with $c' \simeq 0$ will result. Like loose sands, normally consolidated clays can withstand large deformations without rupture. An overconsolidated clay soil will exhibit a fairly linear stress–strain relation and small volume changes until the peak strength is approached. As the peak strength (c' high, $\phi' \simeq 8°$ to $18°$) is attained, a rupture plane will form in the soil and water will be drawn into this plane as the intrinsic

49

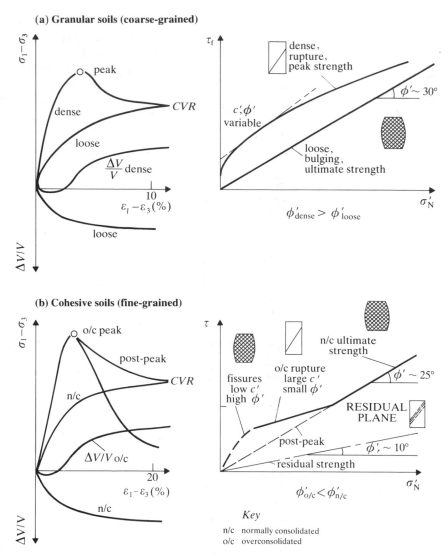

(a) Granular soils (coarse-grained)

(b) Cohesive soils (fine-grained)

Key

n/c normally consolidated
o/c overconsolidated

Figure 2.6 Typical behavior of soils.

bonding breaks down and the soil volume increases toward a critical voids ratio. The planar displacement which develops in laboratory tests promotes an alignment of clay particles (for example, an illite particle magnified 5000 times would look like a razor blade and face-to-face particle contact is easily achieved during planar shear) and the shearing resistance in this plane could be reduced to a residual value ($c' = 0$, ϕ' less than 18° and commonly closer to 8°). When larger particles (kaolin or silt particles) are present in the soil, clay particle alignment is more difficult and the post-peak strength will be

50

greater. In most field situations where the overstressed zone is confined so that a rupture plane does not develop, a general softening will occur in the overstressed zone and the post-peak shearing strength can be as high as the normally consolidated strength under the existing average effective stress. This is often called the fully softened or critical state strength because the soil has softened back to a normally consolidated condition (Roscoe *et al.* 1958, Skempton 1970). A stress–strain curve where a peak strength is reduced to a post-peak strength at small strains is called a brittle stress–strain relation. Brittle materials are susceptible to cracking and should be avoided in the construction of impoundment dams.

Some recent marine and lacustrine sediments exhibit extremely high sensitivities (St = undisturbed vane strength divided by remolded vane strength) due to changes in pore-water chemistry (leaching) or to cementation bonding. These soils are also characterized by a high value of liquidity index (often greater than unity) and a work-softening stress–strain behavior under certain conditions of loading. The behavior of these soils may also be complicated by viscous time effects similar to, but not nearly as pronounced as, the time effects associated with the behavior of frozen soils. Testing of these types of materials for earth structures engineering purposes must be designed with regard to the stress paths that will develop in the field situation, in order to account for variations from the behavior of less sensitive materials.

2.1.5 The effects of intermediate principal stress and anisotropy

While the common triaxial test, in which $\sigma_2' = \sigma_3'$ (compression) or $\sigma_2' = \sigma_1'$ (extension), is generally used to obtain strength data on soils, most field problems have $\sigma_1' > \sigma_2' > \sigma_3'$. Many problems are plane strain problems where there is one long dimension and the intermediate principal stress is approximately midway between the minor and major principal stresses. Plane strain test equipment (and occasionally true triaxial equipment) has been used to research the effects of varying σ_2' on the strength of soils and some typical results are summarized on Table 2.5. It is concluded that the

Table 2.5 Effects of σ_2' on soil strength.

Reference	Soil type	Triaxial data		Plane strain data	
		ϕ'	Cu	ϕ'	Cu
Bishop (1966)	uniform sand	33° loose 42° dense		34° loose 46° dense	
Wade and Henkel (1966)	remolded illite	26°	$0.32\sigma_v'$	27°	$0.28\sigma_v'$
Mitchell and Wong (1973)	sensitive clay	27°	35 to 68 kPa	31.6°	50 to 68 kPa

51

apparent strength parameters (c', ϕ') are generally slightly higher in plane strain tests than in triaxial tests and the relatively simple Mohr–Coulomb equations (in which σ'_2 is ignored) provide marginally conservative values for the strength of earth materials.

Research into the effects of anisotropy has generally been carried out by trimming triaxial samples from block samples of soils such that the sample axis is oriented at different directions to the *in situ* vertical direction (the usual test sample axis direction). A summary of some of this work is given on Table 2.6. It is not difficult to appreciate why varved silts and clays would give a minimum strength when the maximum shear stress (τ_m) or maximum stress ratio (τ_f/σ'_N) is along the orientation of the varves. The concept of strength anisotropy in homogeneous soils is more difficult to understand, but is generally related to an anisotropic pore-water pressure response to loading (i.e. the A parameter in Eq. 2.5 varies with the direction of the major principal stress). Thus it is often found that the undrained strength (Cu) has a directional variation while c' and ϕ' are isotropic. These variations in undrained strength are not generally significant in design, with the noted exceptions of sensitive cemented soils where marked strength anisotropy may be due to residual bonding in the soil structure (Lo & Morin 1972, Wong & Mitchell 1975), and laminated or banded sediments.

Most soil testing apparatuses have fixed directions to the principal applied stresses, while the principal stress directions rotate in a soil mass during construction activities. This limits the direct application of test data to field problems. The simple shear apparatus developed by Roscoe (1953) subjects a soil sample to stress rotation during a shear test and has been used to research the effects of principal stress rotation further (see, for example, Roscoe *et al.* 1967).

Table 2.6 Strength anisotropy.

Reference	Soil type	Undrained strength ratios	
		Cu at $45°/Cu_v$	Cu_h/Cu_v
Hvorslev (1960)	n/c clay	0.92	0.87
	o/c clay	1.08	1.20
Ward *et al.* (1965)	fissured clay	0.93	1.46
Duncan and Seed (1966)	kaolin, $OCR = 9$	0.88	1.16
Lo (1965)	lightly o/c clay	0.90	0.83
Yong and Silvestri (1979)	sensitive clay	0.58	0.69
Lacasse *et al.* (1977)	varved clay	0.5–0.6 $\phi' = 18°$ along varves	0.8 $\phi' = 28°$ across varves

2.1.6   Strength of partly saturated soils

A saturated soil has only one strength envelope and all values of Cu (at various void ratios) are specific points on the c', ϕ' envelope. This is not the case with a partially saturated material since the air pressure and the water pressure can vary independently as volume changes alter the curvature of individual air/water menisci. It is the meniscus tensions that complicate the behavior of partially saturated soils by creating a positive effective stress and an apparent cohesion in the sample when there is no applied external stress. This explains why sand castles, constructed with damp sand, will collapse when allowed to dry or when saturated with water. The apparent cohesion in many natural partly saturated soils is often dangerously transient and increased saturation due to either seepage or volume compression can result in loss of meniscus tensions and collapse of the soil. Indeed, such soils are often referred to as collapsing soils. Total stress strength parameters (c, ϕ) can be obtained from laboratory tests on a partly saturated soil, but it is common practice to saturate such soils by back-pressuring the drainage buret in order to obtain effective stress strength parameters (c', ϕ' as shown on Fig. 2.7). The potential strength loss due to saturation can then be evaluated. The potential for saturation and the risk of strength loss can then be evaluated from site conditions and applied loadings. More fundamental approaches to evaluating the strength of partially saturated soils and experiences with the stability of these soils are contained in publications by Clevenger (1956), Jennings and Knight (1957) and Fredlund (1979).

2.1.7   Cyclic loadings and strength of earth materials

Cyclic stress reversals between compression and extension as well as repeated compressive loadings tend to decrease continuously the volume of particulate materials. The initial response in a saturated material is a continuous build-up of pore-water pressure (Δu) and a decrease in the short-term strength. The classic example of strength loss due to cyclic loading is the liquefaction of loose saturated sands which can develop after a few loading cycles (during an earthquake, pile-driving or blasting operations). Essentially, the excess pore-water pressure increases to approach the average total ground stress ($\Delta u \rightarrow (\sigma_1 + \sigma_3)/2$) such that the strength approaches zero and the material flows. The liquefaction potential (and degree of liquefaction) will depend on the relative density of the sand and the average total stress as well as the level and duration of the cyclic or repeated loading (Seed 1968, Seed & Peacock 1971). Even dense sands can liquefy under high confining stresses combined with a high level of cyclic stress application. A typical liquefaction test result is shown on Figure 2.8. Partially saturated sands can liquefy under continued cyclic loading, but this can be prevented by maintaining a sufficiently low moisture content in the sand that it will not become saturated when densified.

53

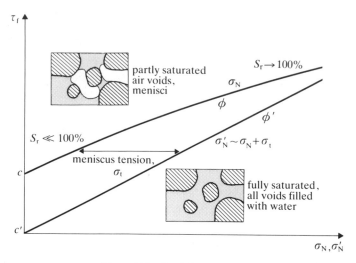

Figure 2.7 Partially saturated soil.

Cohesive soils do not liquefy, but short-term undrained strength reductions of up to 50% have been found in laboratory tests subjected to repeated and cyclic loadings (Seed & Chan 1966, Mitchell & King 1977). Cyclic loadings due to wave action have been found by Mitchell and Hull (1974) to decrease the strength of bottom sediments and promote submarine landsliding in model studies.

In cases where the frequency of the cyclic load is sufficiently low that the excess pore-water pressures dissipate to levels below that required for failure, cyclic or repeated loadings can produce continued settlement (under pavements, track beds and foundations) and can reduce the long-term strength of all earth materials (Larew & Leonards 1962, Brown *et al.* 1975). The reduction in the long-term strength, as observed in laboratory samples where $\Delta u = 0$ at failure, is considered to result due to continued particle degradation or to alignment of particles in shear zones.

2.1.8 Time effects and the strength of earth materials

Granular and select compacted earth materials generally maintain a constant dry density indefinitely and there are no significant time effects once a pore-water pressure equilibrium is achieved. The strength of most soft clay soils tends to increase marginally due to creep consolidation, thixotropic effects (Mitchell & Houston 1969) and weathering. Lightly overconsolidated clays generally remain quite stable, while exposed heavily overconsolidated soils have a tendency to soften due to creep, swelling and weathering. Active clay soils which contain swelling minerals (notably sodium montmorillonite as in bentonitic clays) have a pronounced swelling and an associated decrease

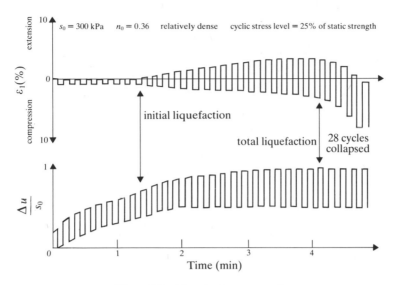

Figure 2.8 Liquefaction test result.

of strength with time when exposed in the presence of free water (Bjerrum 1967b, Kenney 1967a, Jones & Holtz 1973).

Detrimental time effects in most earth materials are associated with stress relief. Soft rocks and very stiff clays will creep soften and can develop weak planes, plastic flow or tensile cracking. Underground openings in fractured rocks are often designed for a limited stand-up time and must be supported or backfilled to prevent closure or collapse (Bieniawski 1974). Weathering along exposed bedding planes or fractures in all types of rocks can result in a long-term decrease of the shearing resistance in these planes (Muller 1964a).

2.1.9 Strength of rock masses

Intact rocks are similar in behavior to plain concrete: strong rocks such as granite, dolerite, basalt and marble exhibit brittle stress–strain relations and will generally burst under unconfined conditions; while weak rocks such as chalk, rock salt and schistotic rocks generally exhibit a more plastic behavior. However, the mass stability of rock exposures depends mainly on the shearing resistance in adversely oriented bedding, folds or fracture planes. Creep strains and weathering on exposure (moisture and temperature changes, chemical oxidation or biological agents) can rapidly reduce joint strength.

Blocky and seamy rocks have sufficiently close joint spacings that the intact strength of the blocks is of secondary concern. Continuity, degree of interlocking and orientation of joint systems, as well as the strength of bedding planes, joints and fractures (surface roughness, infill weathered

55

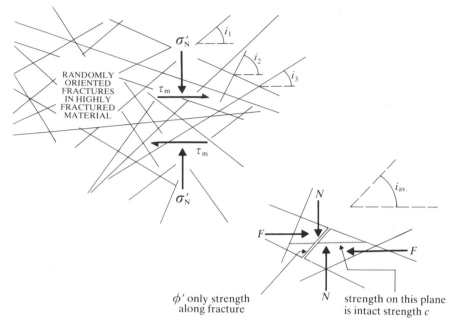

Figure 2.9 Dilative shear mechanism. Material has two limiting modes of failure: (1) failure along the plane of maximum shear or maximum obliquity as is controlled by the properties of the intact pieces or (2) failure by sliding on fracture planes involving a mass dilation and opening of voids within the material giving $\tau_f \sim \sigma'_N \tan(\phi' + i_{avg})$.

material), will be of primary concern (Deere 1964). Detailed geological mapping (joint surveys, structural mapping) and testing of fracture surfaces is necessary to obtain values of c' and ϕ' to be used to analyze various modes of failure (Jaeger 1971).

Fractured and crushed rock can be mass tested in large shear boxes (see, for example, Marsal 1967), or appropriate c', ϕ' values can be estimated from model testing (Ladanyi & Archambault 1970). The strength of this type of material is dependent on the intrinsic strength of rock pieces and the degree of interlocking (particle shape, density), as demonstrated by the dilative shear mechanism outlined on Figure 2.9. Computer modeling of the behavior of fractured rock masses has also been developed (Cundall 1974). Fractured and crushed rock masses are characterized by a small to moderate post-peak decrease in shearing resistance and a strongly curved failure envelope.

Swelling or squeezing rocks are characterized by plastic flow with yield limits that depend on the rate or duration of stress application and the moisture content. Progressive closure or long-term failure can develop in this type of rock. In some cases (salt rock) the creep is also dependent on temperature and the behavior is, in many respects, similar to the behavior of homogeneous frozen soils (McClain 1964, Gouchnour & Andersland 1968).

2.1.10 Stress path effects in material testing

Stress reductions in the absence of failure shear stresses can cause fracturing and fissuring in heavily overconsolidated clay soils and such fissuring can decrease the mass strength of the soil (Lo 1970, Marsland 1971, Singh *et al.* 1973).

Sensitive clays and silty clays, which exist in nature at liquidity indices equal to or greater than unity due to natural cementation or post-depositional leaching of electrolytes from the pore water, exhibit extremely large volume collapse when stresses exceed a yield condition, and it is general practice to maintain foundation loadings on these soils within the yield envelope defined from laboratory shear-consolidation testing (Bjerrum 1967a, Wong & Mitchell 1975). These stiff brittle soils also exhibit closely spaced fissuring which reduces their shearing resistance under low confining stresses and gives a pronounced curvature to the failure envelope (Eden & Mitchell 1970).

The fact that a clay soil can have the two limiting design conditions of short-term (undrained) and long-term (drained) is not surprising. The fact that the same clay soil can have five different drained (long-term) design conditions – normally consolidated, peak overconsolidated, fully softened, residual and fissured – is somewhat confusing. To avoid using the incorrect design condition, the geotechnical engineer should ensure that laboratory or field testing is carried out in such a manner that the loading conditions during the tests are similar to the loading conditions expected in the field. For laboratory triaxial testing it is necessary to differentiate between loading problems (where the average effective stress in the soil is increased during construction) and unloading problems (where the average effective stress in the soil is decreased due to construction). Figure 2.10 considers the two extremes of slope cutting (unloading) and embankment bearing capacity (loading) and shows why slope stability is critical in the long term while bearing capacity is critical in the short term.

Triaxial stress paths can be designed to model field stress paths and particular testing sequences have been recommended for embankment and slope problems (Lambe 1967, Mitchell 1975). Typical stress paths for retaining-wall problems in clay soils have been discussed in research papers (Henkel 1970a, McRostie *et al.* 1972) and this type of elemental model testing confirms field observations that the strains (and strengths) in soils can be influenced by the methods and rate of construction. Although many foundation loadings produce stress paths that are suitably modeled by the standard triaxial tests, slope unloadings and large embankment loadings may require testing outside of the normal triaxial test range (see Fig. 2.3).

57

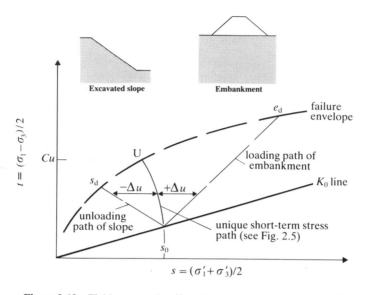

Figure 2.10 Field stress paths. *Slope:* long-term strength at point s_d is less than the short-term strength at point U; the strength decreases with time as the pore-water suctions $(-\Delta u)$ dissipate. *Embankment:* short-term strength at point U is less than the long-term strength at point e_d; the bearing capacity is critical in the short term and increases with time as $+\Delta u$ dissipates.

Strength test data must not be extrapolated beyond the range where test points are obtained.

Many geotechnical engineers also support the concept of using different field tests to determine soil properties for different problems, although this is not always possible. It is also noted that most of the common field tests are more applicable to foundation engineering than to earth structures engineering. Earth structures engineering relies mainly on sampling and laboratory test techniques to obtain design strength criterion. Large-scale field tests or model studies are also used in earth structures engineering (Krsmanovic & Popovic 1966, Kwan 1970, Schofield 1978), and *post-mortem* analysis of field failures are most valuable in evaluating testing and design procedures.

2.2 Ground water and earth structures

Ground water is a major concern of earth structures engineering. For major projects both regional and local groundwater conditions must be considered. Existing well records can be interpreted with reference to topography and soil and rock conditions to provide information on regional groundwater

Table 2.7   Typical permeabilities of earth materials.

Soil type	Typical natural permeability (m s^{-1})	Factors controlling permeability	Empirical relations (k in m s^{-1} D_{10} in mm)	Methods of decreasing permeability
clean gravels	10^{-1}–1	particle shape, density	none	cement grout
uniform coarse sand	10^{-2}–10^{-1}	void ratio, saturation	$k = 0.03 \left(\dfrac{D_{10}e}{0.85}\right)^2$	silicates, densification
uniform medium to fine sand	10^{-5}–10^{-3}	void ratio, saturation	$k = 0.015 \left(\dfrac{D_{10}e}{0.85}\right)^2$	emulsions, densification
well graded sands	10^{-4}–10^{-2}	void ratio, saturation	—	emulsions, densification
fissured clays	10^{-5}–10^{-3}	saturation fissuring	—	cement grout
silts and clay silts	10^{-7}–10^{-5}	void ratio, grain sizes	—	resin grouts
varved clays	10^{-6}–10^{-5}	void ratios	$k_h = 10$ to 100 times k_v	resin grouts
non-plastic clay	10^{-7}–10^{-8}	void ratio, mineralogy	—	electrochemical
plastic clay	10^{-10}–10^{-8}	plasticity, mineralogy	—	electrochemical
intact and massive rock	10^{-14}–10^{-10}	joint spacing, cavities	—	—
blocky rock	10^{-8}–10^{-6}	joint spacing, cavities	—	cement grout
fractured rock	10^{-3}–10^{-1}	weathering products	—	cement grout

Non-fissured fine-grained soils assumed to be saturated.

aquifers. Piezometer measurements of groundwater pressures and pressure changes during well pumping trials can provide information on the hydraulic conductivity of subsurface soils and rocks and on the connectivity of aquifers. Field and laboratory measurements of permeability (hydraulic conductivity) of earth materials provide data for analyses of how the local groundwater flow regime will be altered by proposed construction activities. Correct interpretation of flow regimes and correct evaluation of construction effects requires a knowledge of the factors affecting groundwater flow.

2.2.1   Soil moisture and ground water

Water accumulating on the surface of pervious soils or rocks will percolate into the ground under the influence of gravity, join with the existing ground water and flow laterally toward major surface or subsurface drainage sinks (rivers, lakes, rock faults, buried pervious channels). This causes seasonal changes in the level of the groundwater table depending mainly on climatic conditions. The quantity of groundwater flow depends on the permeability of the earth materials and the hydraulic gradient ($Q = kiA$). In rocks water travels along joint systems, and intact rocks can be sufficiently impervious that the water is connate. Table 2.7 lists typical permeabilities of various earth materials and some factors affecting permeability.

Lateral flow gradients generally conform regionally to the topographical slopes but considerable local variations can be expected. Vertical gradients are created by subsurface aquifers or sinks which are generally confined zones of pervious materials in which the water pressure is different from hydrostatic (hydrostatic pressure $u_h = h_w\gamma_w$ where h_w is the depth below the water table). Figure 2.11 shows three idealized conditions of vertical hydraulic gradients. In some cases an aquifer may be separated from the near-surface ground water by a relatively impervious layer and there is little vertical flow. Thus the connectivity is low and the near-surface water is perched on the impervious layer. Creating a drainage sink at depth would have little effect in this case. In other cases the creation of a subsurface opening would initiate a drawdown condition with the possibility of depleting surface waters and

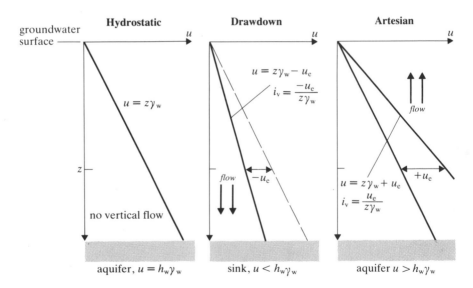

Figure 2.11   Idealized vertical flow gradients.

60

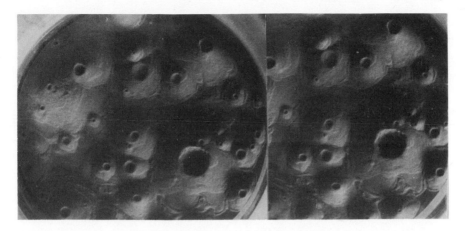

Figure 2.12   Sand boils produced in the laboratory.

creating ground subsidence. Such conditions are considered in more detail in subsequent chapters.

Artesian pressures develop in nature in low-lying sandy or silty areas where there is adjacent free water at a higher elevation or where there is a confined aquifer. In some cases the upward gradient can approach a critical value ($i_c = \gamma'/\gamma_w$) such that $\sigma'_v = 0$. This condition is known as quicksand and the soil mass will increase in volume and will be floated by the upward seepage stresses (drag forces/area). The soil mass will then have the characteristics of a dense fluid and will not float a mass of density greater than γ. In extreme cases vertical gradients can exceed i_c and turbulent flow with open pipes will develop in the soil; soil particles will be transported up the flow pipes forming conical mounds or boils on the ground surface as shown on Figure 2.12. Critical exit gradients and soil piping are major problems during excavations in saturated sandy soils. Avoiding piping failures is a major design consideration in earth impoundment structures such as dams and levees. An exit gradient is defined as the hydraulic gradient at a point where water is flowing out of a soil mass, and the value $i_c = \gamma'/\gamma_w$ is only applicable for the vertical component of the exit hydraulic gradient. Where flow is exiting on a sloped surface, the critical exit gradient can be approximated as

$$i_c = (\gamma'/\gamma_w) \cos \beta$$

as shown on Figure 2.13. In the case of near-horizontal flow from a vertical exposed face, the critical gradient approaches zero and a continued erosion of the face can develop as the exit gradients remove soil particles – this process is referred to as internal erosion or sapping and can take place very slowly or very rapidly. Flow gradients are often crucial in determining the

61

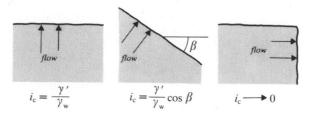

$$i_c = \frac{\gamma'}{\gamma_w} \qquad i_c = \frac{\gamma'}{\gamma_w} \cos \beta \qquad i_c \longrightarrow 0$$

Figure 2.13 Critical exit gradients.

stability of earth structures and will be considered again in subsequent chapters.

Soil drainage is carried out using pumped wells to facilitate excavation in cohesionless soils, using various types of drains to improve the stability of earth structures or to speed up the consolidation process, and using trenches or gravity drains to reduce the water content of borrow materials for compaction. As shown on Figure 2.14, drainage leaves capillary water in a soil with the degree of saturation depending mainly on the grain size, uniformity and relative density of the material. For uniform sands and silts the height of capillary rise may be estimated as $h_c \approx 0.06/eD_{10}$ meters, where e is the void ratio and D_{10} is the effective grain size in millimeters. The capillary rise in very fine-grained soils can exceed 30 m. Some or all of this water can be removed by desiccation drying or by electro-osmosis, and such procedures are occasionally necessary in geotechnical engineering. The movement of capillary water through soils is of special interest in considerations of frost action and contaminant migration in soils.

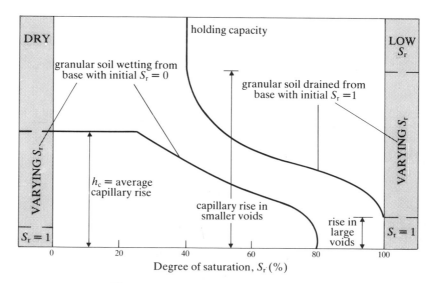

Figure 2.14 Capillary heads in soil columns.

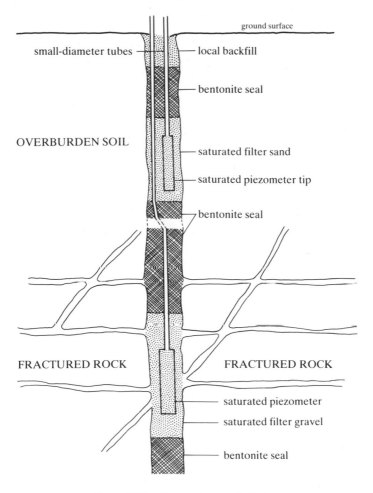

Figure 2.15 Piezometer installations.

2.2.2 *Groundwater pressures and field permeability tests*

The mass permeability or hydraulic conductivity of a porous material like soil or rock is defined as the velocity at which water will flow through that material under a hydraulic gradient of unity. Velocity is defined as $v = Q/A$ where A is the total soil area perpendicular to the flow. The permeability of earth materials is most accurately measured by field tests and these tests can be carried out with the same field instruments used to measure the ground-water pressures. The same basic techniques can be used to measure water pressures and *in-situ* permeabilities in soils and rocks.

For water pressure measurements, piezometer tips should be sealed into the stratigraphic or lithographic unit, ensuring good communication with

63

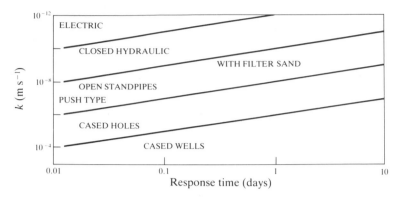

Figure 2.16 Piezometer and well response.

the ground water (see Fig. 2.15). In granular soils and fractured rock there is little problem with water communication to the tip. In cohesive soils, the cavity in which the piezometer tip is to be located should be wash-bored below the casing to eliminate surface smear and the tip should be surrounded with clean filter sand. In more competent rocks the tip length should be sufficient to intersect water-bearing features (two or three times the joint spacing). There are a variety of piezometer tips available of porous stone or porous plastic construction that act like a well point in that they allow free passage of ground water while excluding soil particles. Some tips are designed to be pushed into soft soils (attached to a steel drive pipe). These are most useful for regional groundwater studies because they do not require costly boring and sealing but they do cause surface smear and have a slower response time.

Response time, for a piezometer, is the time required for the instrument to register a rapid change in groundwater pressure due, for example, to infiltration of rainfall or a drawdown caused by a tunnel or a mine opening. This is also referred to as lag time. The response time will depend mainly on the size of the tip and the compliance of the measuring system relative to the permeability of the soil (the compliance being the volume of water that must flow into or out of the tip to register unit water pressure change). Large-compliance piezometers are those with open standpipes attached to the tip (see Fig. 2.15) and the water level is measured using a dip-wire with exposed points which close a d.c. circuit when water contact is made. Small-compliance piezometers are those with pressure transducers (electrical or hydraulic) located in the tip and suitable leads to activate and monitor the transducer. A typical compliance for this latter type of piezometer is about 10^{-6} cm^3 per meter of head change. Figure 2.16 can be used as a guide for selection of piezometer compliance for different response requirements. All types of tips should be de-aired (saturated) and placed in a saturated soil for groundwater pressure measurements. There are special ceramic tips avail-

64

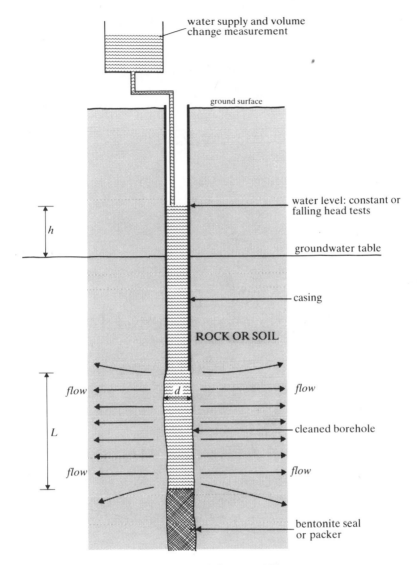

water supply and volume change measurement

ground surface

water level: constant or falling head tests

h

groundwater table

casing

ROCK OR SOIL

flow $\qquad$ d $\qquad$ flow

L

cleaned borehole

flow $\qquad$ flow

bentonite seal or packer

Figure 2.17 Borehole permeability tests.

able for measurement of soil suction in partially saturated soils but these are of little general interest in earth structures engineering. Several piezometers sealed at different depths in one borehole are referred to as a string of piezometers or a piezometer string.

To conduct field permeability measurements one must create a local groundwater flow condition and monitor the time rate of flow. This can be accomplished in a cased hole by isolating a test section as shown on Figure

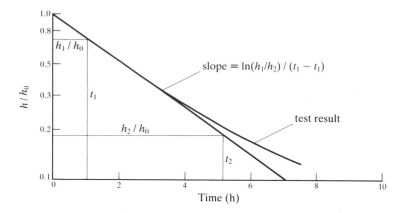

Figure 2.18   Falling-head test data.

2.17. The length of the test section (L) can be as long as practicable from stratigraphic, borehole stability and flow volume recording considerations and should be at least four times the diameter (d). The natural groundwater level is measured in the casing and a small head (h) is created and maintained until the flow regime is established. In a saturated material this takes only a few minutes, but it may take several hours to saturate an unsaturated material. The head should be kept small when possible (less than 5 m) to avoid leakage around the casing.

Two types of tests are commonly used:

(a) In a constant-head test the casing water level is maintained by controlling the flow (manual adjustment or servo-valves operated by contact electrodes or floats can be employed) and the flow Q is measured as the volume of water flow per unit time. The flow quantity is given by $Q = kiA$, where A and i are variable with distance from the borehole. The flow quantities can be integrated over the zone of influence to give

$$k = \frac{Q \ln(2mL/d)}{2\pi Lh} \qquad (2.10)$$

where m is equal to $\sqrt{(k_h/k_v)}$ and is unity for an isotropic material. If the test section is located directly below an impervious boundary, the value of m is doubled.

(b) In a falling-head test the casing is filled to an initial level and the head loss with time is monitored using a level indicator probe. The results are plotted on a graph as shown on Figure 2.18 and the initial tangent slope is evaluated as

$$a = \ln(h_1/h_2)/(t_2 - t_1) \qquad (2.11)$$

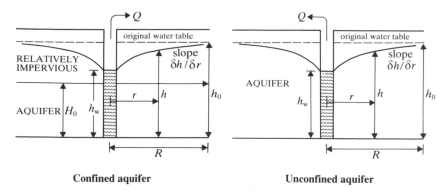

Confined aquifer   **Unconfined aquifer**

Figure 2.19   Radial flow to wells.

The mean coefficient of permeability is then given as

$$k = \frac{d^2 a \ln(2mL/d)}{8L} \tag{2.12}$$

With the falling-head test it is only necessary to be able to raise the water level in the casing and accurately time the rate of head loss. A long displacement float and a watch are the basic tools. This is the reason why falling-head tests are usually preferred in open boreholes.

Piezometer installations can be used for *in-situ* permeability measurement and offer the advantage of being permanent installations where the effect of construction on the material permeability can be monitored from time to time. In slopes or mine crown pillars, for example, the effects of small ground movements (subsidence) on ground permeability can be monitored.

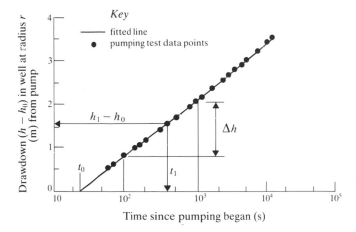

Figure 2.20   Typical drawdown relation.

Using piezometer installations the values of L and d are taken as the dimensions of the filter material surrounding the tip.

2.2.3   Well pumping trials for aquifer permeability

Permeability tests measure local permeabilities. A large-scale well pumping test, where water is pumped out of a borehole and water levels are monitored in observation wells (or piezometer installations) across the site, is used to evaluate the mass (or average) aquifer parameters. Two basic stratigraphic conditions are considered as shown on Figure 2.19. For steady-state flow where the well is being pumped at a constant rate Q and the phreatic surface has stabilized in observation wells and in the pumping well, the mass horizontal coefficient of permeability is

$$k = Q \ln(r_2/r_1) / 2\pi H_0(h_2 - h_1) \qquad \text{for a confined aquifer} \qquad (2.13)$$

$$k = Q \ln(r_2/r_1) / \pi(h_2^2 - h_1^2) \qquad \text{for an unconfined aquifer} \qquad (2.14)$$

For solution, the water levels h_1 and h_2 in observation wells (or piezometer stations) at radii r_1 and r_2 from the pumping well should be known. If only one observation well is available, the borehole radius and h_w may be substituted for r_1 and h_1 with some loss of accuracy. If no observation wells are available, $\ln(r_2/r_1)$ may be assumed to be $\ln 100 = 4.6$, and h_2 is equated to h_0 (the original groundwater level) to obtain an approximate answer.

For unsteady-state flow the well should be pumped for several hours with piezometer observations of drawdown in close proximity to the well (less than 100 m). Following the solution due to Jacob (1950) the pumping test data are plotted as shown on Figure 2.20 and the permeability coefficient for a confined aquifer is estimated as

$$k = 0.18Q/\Delta h(H_0) \qquad (2.15)$$

Equation 2.15 can be used for an unconfined aquifer providing that the drawdown is small compared to the aquifer thickness – otherwise the quantity $(h_0^2 - h^2)/2h_0$ should be plotted instead of $(h_0 - h)$ on Figure 2.20.

The specific yield or storage coefficient (the volume of water stored per unit surface area of the aquifer per unit change in the water head normal to that surface) is estimated as

$$S = 0.05kH_0t_0/r^2 \qquad (2.16)$$

Other methods of solution for unsteady-state flow include the Theis method which uses a match point curve and the Chow method which uses a nomograph. These methods are outlined in groundwater hydrology texts.

CONFINED FLOW
Wall

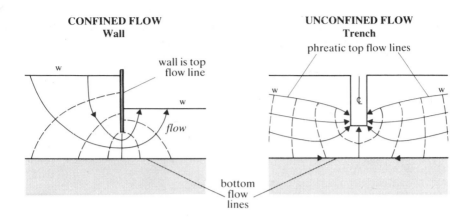

wall is top flow line

w

w

flow

bottom flow lines

UNCONFINED FLOW
Trench

phreatic top flow lines

w

w

Weir

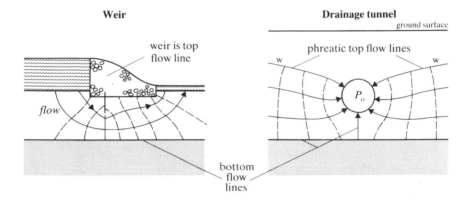

weir is top flow line

flow

bottom flow lines

Drainage tunnel

ground surface

phreatic top flow lines

w

w

P_0

Retaining wall with base filter

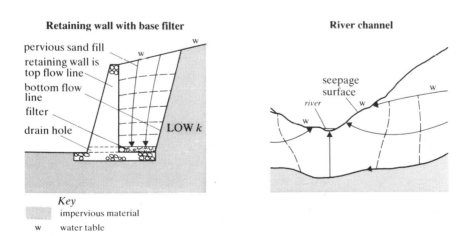

pervious sand fill

w

w

retaining wall is top flow line

bottom flow line

filter

drain hole

LOW k

River channel

seepage surface

river

w

w

w

Key

impervious material

w   water table

Figure 2.21   Flow net examples.

Horizontal anisotropy in the aquifer (due, for example, to fault zones in rocks) can be discovered by having observation wells or piezometer stations located at various radii in a number of directions (in four directions at right angles, for example). Anisotropy would be indicated by different drawdown curves in different directions. If sufficient observations are available the drawdown can be contoured as equipotentials. Circular plan equipotentials would indicate isotropy and elliptically shaped contours would define major and minor axes with the permeability in these directions being inversely proportional to the ratio of the axis length.

Connectivity between the aquifer and adjacent ground water at lesser or greater depths can be evaluated by monitoring the effects of drawdown on piezometric levels in these adjacent materials (see, for example, Frind 1970, Debidin & Lee 1980).

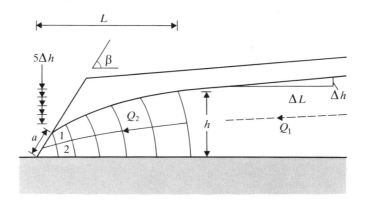

$$Q_1 = kiA = k\frac{\Delta h}{\Delta L} \quad h = Q_2 = kh\frac{N_f}{N_d} \quad \begin{array}{l} N_f = 2 \\ N_d = 6 \end{array}$$

$$\text{also } Q_2 \approx k\,[\,h^2 - (a\sin\beta)^2\,]/L$$

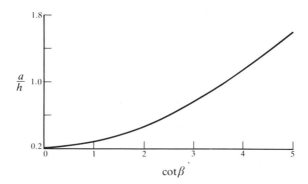

Figure 2.22 Seepage and drawdown in excavations.

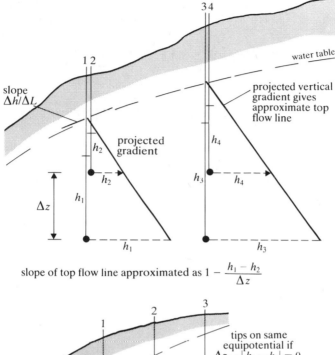

Figure 2.23   Graphical techniques for piezometric data interpretation.

2.2.4   Flow nets for earth structures

Figure 2.21 shows several examples of confined (top flow line formed by impervious material) and unconfined (top flow line is phreatic) steady-state flow nets. These are drawn by methods outlined in basic soil mechanics texts. Equilibrium flow nets for excavation slopes in homogenous soils can be estimated as outlined on Figure 2.22 but the flow regime in natural slopes should always be evaluated from piezometer installations. Some simple graphical techniques for establishing a flow net in a homogeneous material from piezometer data are illustrated on Figure 2.23. Figure 2.24 outlines the

71

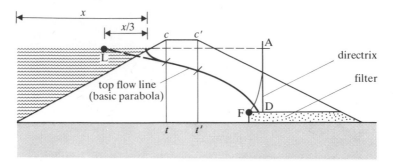

Figure 2.24 Top flow line construction (scaled section). The upper interior corner of the filter is the focus of the basic parabola and a point L is obtained as $x/3$ as shown. The directrix of the parabola is obtained from the arc FA (with centre L) and the extension of the water level LA. Points on the parabola are obtained from the condition that they are equidistant from the directrix and focus. The top flow line is coincident with the basic parabola but diverges to meet the reservoir level on the upstream face.

construction of a phreatic flow line in a homogeneous earth dam with a base filter.

Basic soil mechanics texts show the development of a transformation method as outlined on Figure 2.25 for drawing flow nets in anisotropic or layered systems, but more sophisticated analytical methods may be required for evaluating unconfined flow in non-homogeneous soils, particularly when regional flow conditions are being evaluated (Hodge & Freeze 1977, LaFleur & Lefebvre 1980). Basic texts also outline graphical transfer methods for determining the change in gradient when water flows from one soil type to another, but detailed flow calculations in earth structures may either require numerical techniques (see, for example, Finn 1967) or the permeability of the materials may be sufficiently different that the flow net may be restricted to the least pervious material. Regional groundwater flow analysis is generally quite complex, involving the effects of precipitation, infiltration and connected aquifers. The flow into unit length of a temporary excavation can be estimated from the formula

$$Q = kh/(4 + \cot \beta) \qquad (2.17)$$

which is obtained by using $L = h(4 + \cot \beta)$ in the approximate expression for Q_2 on Figure 2.22. The use of flow nets in stability analysis of earth structures is discussed further in Chapters 4 and 5.

2.2.5 Subsurface drainage and groundwater control

For construction de-watering purposes, gravity or suction wells are often used. Major de-watering schemes can often produce construction economies

True scaled section

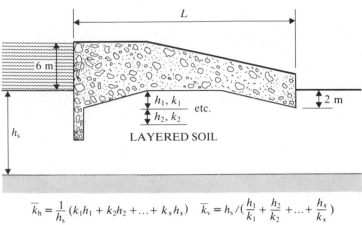

$$\overline{k}_h = \frac{1}{h_s}(k_1 h_1 + k_2 h_2 + \ldots + k_x h_x) \quad \overline{k}_v = h_s / (\frac{h_1}{k_1} + \frac{h_2}{k_2} + \ldots + \frac{h_x}{k_x})$$

Transformed section

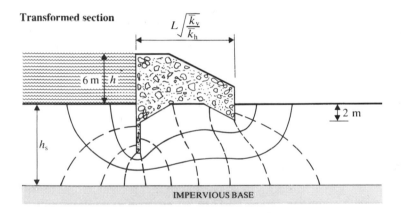

Figure 2.25 Flow nets for layered soils. Draw flow net on transformed section as shown and complete calculations as usual except $Q = (\overline{k}_h \overline{k}_v)^{1/2} h N_f / N_d$. A similar transformation is used when the soil is not isotropic with respect to permeability or when there is a variation in permeability with depth. In all cases the value of $\overline{k}_h$ will be greater than the value of $\overline{k}_v$.

for large earthworks by improving both the stability of slopes and the working conditions for excavation equipment. Figure 2.26 shows earth-moving equipment working at a depth of about 8 m below the original water table in a soft silty clay. The water table was lowered to several meters below the bottom of the excavation in order to make excavation by scrapers possible. Large-diameter gravity wells with well pumps are most common for major de-watering and can be used in conjunction with recharge wells in order to reduce the regional influence and keep the major effects of consolidation subsidence to the area between the de-watering and recharge wells.

73

Figure 2.26   De-watering to facilitate construction.

Figure 2.27a shows a typical arrangement of wells for major de-watering purposes. Grout injection or sheet pile cutoffs may be used to isolate aquifers adjacent to major de-watering installations.

Local de-watering is often necessary in conjunction with trench excavations or temporary retaining walls in saturated granular materials. Well points connected to a suction header are usually jetted into place with a filter sand or gravel surround to increase the zone of influence and prevent fine particles from being pumped out. The main purpose of suction wells is to intercept local flow to prevent loss of ground due to piping, but such systems can also provide temporary strengthening of a frictional soil by reducing the pore-water pressure below the hydrostatic value. Mansur and Kaufman (1962) provide a nomograph for well point spacings and a typical well point system is shown on Figure 2.27b. Eductor well points can be used for depths up to about 30 m but are quite expensive compared to simple suction well point systems. Grouting to decrease the permeability of a material and provide some cohesion between particles may be used in some situations as an alternative to well point de-watering. Figure 2.28 indicates soil grain sizes for which various treatments are applicable. Drainage systems to increase consolidation rates or to effect a permanent groundwater pressure reduction in slopes are discussed in subsequent chapters and foundation grouting for earth dams is discussed in Chapter 5.

2.3   Settlement of earth structures

Soil and rock construction materials are compressible but the compression generally decreases as grain size or density increases. The density of earth

(a)

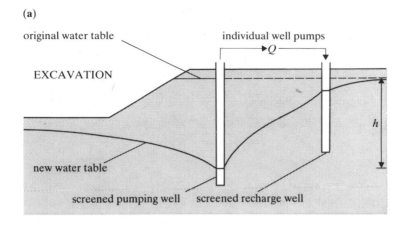

(b)

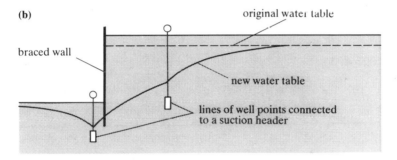

Figure 2.27 Typical de-watering systems. (a) Major de-watering with recharge wells. Typical well spacings in the order of $2h$ to $5h$. (b) Construction de-watering using well points at typical spacings in the order of 1 m to 3 m.

construction materials is increased by compaction, and densification can sometimes be achieved in natural foundation materials. This section reviews the basic concepts of compaction, compression and settlement. Particular design and performance applications of settlement analysis to earth structures are presented in later chapters.

2.3.1 Compaction and compression

Compaction of an earth material is defined as the densification of that material under transient or dynamic loads. The Proctor compaction tests (the standard test is designed to correlate with standard field compaction equipment and the modified test is designed to correlate with high-energy field compactors) are laboratory control tests that are carried out to see what densities can be achieved in field placement and compaction of earth materials. Specifications based on these laboratory tests are checked in the

75

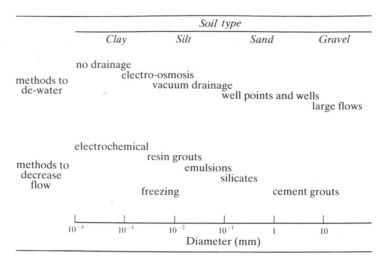

Figure 2.28 De-watering and grouting.

field using various field density test methods (see, for example, Bowles 1978).

The dry density achieved in a compaction test is found to depend on the compaction moisture content and an optimum moisture content (w_{opt}) is defined when the maximum density ($\rho_d(max)$) is achieved for a given compaction effort. Specifications require the contractor to achieve a certain percentage of the maximum dry density in earthworks (usually between 90 and 100% depending on the type of earthwork and the test used). It is of economic advantage to the contractor to compact at the optimum water content because the specified dry density can be achieved with less work. In some cases a compaction water content may be specified (this is usually done to prevent overcompaction which could produce a very dense brittle material where a more resilient material may be desired) with reference to the zero air voids line which represents the maximum density that can be achieved by transient loads. Some typical characteristics and properties of compacted earth materials are given on Table 2.8.

Compressibility of an earth material is defined by the volume compression that will develop under static loads. To test for soil compression characteristics, an undisturbed sample is contained in a thick-walled metal ring and subjected to a vertical (one-dimensional) loading using a piston that fits into the ring. The void ratio at pore-water pressure equilibrium under each loading increment is plotted against the vertical effective stress in the soil, as shown on Figure 2.29. The time required to reach pore-water pressure equilibrium will vary from a few minutes for clean sands to several hours for saturated impervious clays due to the fact that pore water must be expelled from the sample to accommodate the volume compression. This hydro-

Table 2.8 Typical properties of compacted materials.

Material	Value for embankments	k (m s^{-1})	Compaction seepage	γ_d (kN m^{-3})	w_{opt} (%)	Typical strength	
						c' (kPa)	ϕ' (deg)
crushed rock fill	very stable, pervious	10^{-1}	heavy vibratory, positive cutoff	18–21	—	0	45
gravels GW, GP	very stable	10^{-4}	vibratory roller, positive cutoff	18–22	8–12	0	37
gravels GM, GC	reasonably stable	10^{-5} 10^{-8}	rubber-tired sheepsfoot	18–22	10–14	0 5	34 31
sands SW, SP	stable, unsaturated	10^{-5}	rubber-tired tractor	16–20	9–18	0	37
sands SM, SC	fairly stable, dam cores	10^{-5} 10^{-8}	rubber-tired sheepsfoot	17–20	11–20	10 20	31 34
silts ML, MH	poor stability	10^{-6} 10^{-8}	rubber-tired sheepsfoot	12–19	15–35	10 20	32 25
clays CL, CH	fairly stable, dam cores	10^{-7} 10^{-9}	sheepsfoot roller	14–20	15–35	15 15	20 28

Organic materials are not suitable for embankments and should be removed.

dynamic time delay is called consolidation and is discussed in the following section.

A clay soil will generally show a distinct break in the void ratio versus applied stress relationship and, following the break, the extended relationship will plot as a straight line on the semi-logarithmic plot shown on Figure 2.29. The logarithmic slope (C_c) is defined as the compression index. An unloading–reloading cycle will define an average slope (C_r = swelling or recompression index) as shown on Figure 2.29, and the break defining structural yielding of the soil (point y) will occur at the stress level where unloading was initiated (at the previous maximum stress level to which the soil was subjected). The initial break in the relationship for an undisturbed soil can represent the maximum vertical stress to which the soil had been subjected during its geological history and is defined, on Figure 2.29, as the preconsolidation pressure, p_c'. At stresses less than p_c' the soil is called overconsolidated, and it is termed normally consolidated when the stress level exceeds p_c'. True overconsolidation can be caused by glacial overriding, erosion of previous overburden, excavation or groundwater level changes, but an apparent p_c' can be created by physicochemical effects (syneresis, leaching or cementation).

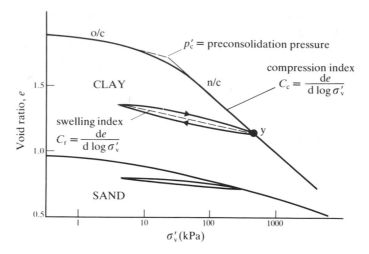

Figure 2.29 Compression curves.

A cohesionless soil such as sand generally shows no distinct break but is not elastic (the unloading–reloading cycle does not follow the original loading relation). The compression relation of a sand depends mainly on its initial relative density, but at some stress level a straight line approximation will be achieved on the semi-logarithmic plot of void ratio versus stress. Typical compression indices (C_c values) for various natural earth materials are noted on Table 2.9 and C_r is commonly about 0.2 C_c.

When stress levels are in excess of p_c' the vertical strain in the sample is given by

$$\frac{\Delta H}{H} = \epsilon_1 = \frac{C_c}{1+e_0} \log \left(1 + \frac{\Delta \sigma_v'}{\sigma_0'}\right) \tag{2.18}$$

where e_0 and σ_0' represent the initial conditions and $\Delta \sigma_v'$ is the effective stress increase. The index C_r may be substituted for C_c during an unloading–reloading stress cycle.

When stress levels are below p_c' it is recommended that the strain be represented by a quasi-elastic equation in the form

$$\frac{\Delta H}{H} = \epsilon_1 = m_v \Delta \sigma_v' \tag{2.19}$$

where

$$m_v = \Delta e / \Delta \sigma_v' (1 + e_0)$$

78

is called the coefficient of volume compression and is obtained from the test relation over the desired stress increment and is the inverse of the constrained modulus in elastic theory. Often the initial portion of the relation can be approximated as a straight line in a linear plot of e versus σ'_v and m_v is then a constant quantity. Unloading–reloading cycles can also be approximated using the quasi-elastic approach. Typical values of m_v are noted on Table 2.9.

When an applied effective stress ($\Delta\sigma'_v$) increases the stress level from an initial value of $\sigma'_0 < p'_c$ to a final value of $\sigma'_f > p'_c$ the strain calculation (or settlement calculation) is divided into two parts as

$$\Delta H = H \left[m_v(p'_c - \sigma'_0) + \frac{C_c}{1 + e_0} \log\left(\sigma'_f/p'_c \right) \right] \qquad (2.20)$$

Table 2.9 Typical strength and compression characteristics of earth materials.

Material	Unconfined $[\sigma_1]_f$(kPa)	Effective		Natural compression		Compacted m_v(kPa^{-1})
		c' (kPa)	ϕ' (deg)	m_v(kPa^{-1})	C_c	
rock, hard	$>10^5$	highly		10^{-7}–10^{-8}	—	—
soft	$<5 \times 10^4$	variable		10^{-6}–10^{-7}	—	—
crushed	0	0	>45	10^{-4}	0.01	0.3×10^{-4}
gravels, low fines	0	0	35 45	10^{-5}	0.01	0.3×10^{-4} 0.6×10^{-4}
gravels, high fines	5	5	30 35	10^{-4}	0.03 0.12	0.5×10^{-4} 1.1×10^{-4}
sands, low fines	0	0	30 45	10^{-5}	0.01 0.08	0.4×10^{-4} 1.0×10^{-4}
sands, high fines	10	5	30 35	10^{-4}	0.03 0.15	0.5×10^{-4} 1.5×10^{-4}
silts, low plasticity	10	0	25 35	10^{-4}	0.03 0.15	0.6×10^{-4} 1.2×10^{-4}
silts, high plasticity	25	10	22 28	10^{-3}	0.3 $0.3(e_0 - 0.3)$	1.4×10^{-4} 2.5×10^{-4}
soft to medium clay	<100	0	22 28	10^{-3}	0.3–0.4 $0.85w^{1.5}$	0.8×10^{-4} to
medium to stiff clay	100–800	50 can soften	<20	10^{-4}	$0.01(w_L - 0.1)$ 0.3–0.4	2.6×10^{-4} depending
sensitive silty clays	25–100	10	30	10^{-4}	0.5–1.0 $0.7(e_0 - 0.5)$	on plasticity

The natural properties of coarse-grained soils depend mainly on relative density and particle shape.
Empirical relations for C_c of fine-grained soils are detailed by Azzouz et al. (1976).
Natural values of m_v are highly variable and only the order of magnitude is indicated.

Most cohesive soils exhibit some creep strains that continue indefinitely (sometimes called secondary compression). These have generally been found to decrease exponentially with time such that a plot of ϵ_1 versus time on a log scale (following pore-water pressure equilibrium at $t = t_{100}$) produces a straight line with log slope $C_\alpha = \epsilon_1$ per log cycle of time. Such creep settlements may be added to the calculated settlements as

$$\Delta H_\alpha = HC_\alpha \log(t/t_{100})$$

to provide a continued settlement prediction over the life of a structure.

While there is no direct general correlation between compression (static loading) and compaction (dynamic or transient loading), it is known that a well compacted soil will exhibit less settlement than the same soil poorly compacted. Densification will also reduce, but not eliminate, the risk of liquefaction in saturated cohesionless soils.

It should be noted that, once a soil is compacted to a specified dry density (or dry unit weight), dry density will not change seasonally unless the soil contains swelling minerals or is subjected to frost action. The bulk (total) unit weight, at any time, is given by

$$\gamma = \gamma_d(1 + w) \tag{2.21}$$

2.3.2 Consolidation and settlement of earth structures

Consolidation is the hydrodynamic process by which water is expelled from saturated soil voids. Fine-grained soils have sufficiently low permeability that water cannot readily escape from the pore space when the grain structure is compressed. Thus, when a load is applied rapidly in a compression test, an initial excess pore water pressure (Δu) equal to the applied load is generated. This excess pressure will initiate flow of water to the sample boundaries and Δu will decrease toward zero as the grain structure compresses. It is of interest to geotechnical engineers to be able to predict the time required for this process to be completed in a field situation. To do this the one-dimensional consolidation theory developed by Terzaghi (1943) is generally used to extrapolate laboratory test data to the field scale. In this respect the consolidation test is a one-dimensional model test and can be subject to scaling errors.

The derivation of the Terzaghi consolidation equation can be found in most basic soil mechanics texts and is outlined in detail by Terzaghi and Peck (1967). The equation assumes linear compression (m_v) and that the rate of decrease of pore-water pressure ($-\partial u/\partial t$) is equal to the rate of increase in effective stress ($\partial \sigma_v'/\partial t$) and is given as

$$\frac{\partial u}{\partial t} = \frac{k}{\gamma_w m_v} \frac{\partial^2 u}{\partial z^2} = C_v \frac{\partial^2 u}{\partial z^2} \tag{2.22}$$

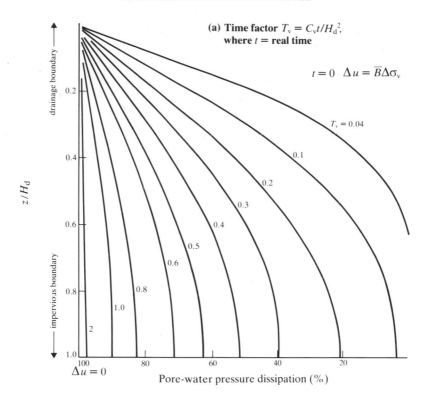

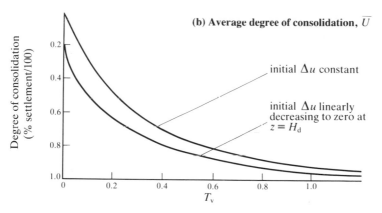

Figure 2.30 Consolidation solutions.

where z is the distance from a drainage boundary and C_v is called the coefficient of consolidation. The general solution of this equation produces a family of curves representing various dimensionless time factors,

$$T_v = (c_v/H_d^2)t$$

81

where H_d is the maximum drainage path length and t is the real time. The solution for Δu = constant at $t = t_0$ is plotted on Figure 2.30a. At any given time factor the average degree of consolidation is defined as the area above that time factor curve divided by the total area in the block diagram of Figure 2.30a. From this definition a curve of the average degree of consolidation versus time factor can be obtained and this curve is plotted on Figure 2.30b. In cases where the compressible soil is deep with respect to the width of the loaded area, the initial excess pore-water pressures are better approximated by a triangular distribution, decreasing with depth rather than constant as shown on Figure 2.30a and the average degree of consolidation for this case is also shown on Figure 2.30b.

Many non-linear consolidation theories have been developed since this original development by Terzaghi. Some theories attempt to account for the non-linear compression characteristic of soils and others attempt to include soil viscosity as a time effect, particularly for sensitive soils that often exhibit large creep (secondary consolidation) rates (Barden 1965, Davis & Raymond 1965). Practicing engineers generally use the original Terzaghi approach, however, and it is considered adequate for most earth structures problems. Although it is known that m_v decreases with effective stress increase, the coefficient of permeability also decreases such that the value of C_v can remain fairly constant over a range of applied stress. Nevertheless, it is recommended that laboratory testing be carried out in the stress range applicable to the field problem. It is also recognized that C_v can decrease markedly as the applied stresses attain and exceed the apparent preconsolidation pressure in some sensitive clay soils.

In practice it can generally be assumed that consolidation is complete ($\Delta u \rightarrow 0$) at $T_v = 0.85$. This corresponds closely to a theoretical degree of consolidation of 90%. The field consolidation time is then related to the laboratory time as

$$t_{90}(\text{field}) = t_{90}(\text{lab}) \frac{H_d^2(\text{field})}{H_d^2(\text{lab})} \qquad (2.23)$$

where $t_{90}(\text{lab})$ is obtained from a root time fitting method which can be found in most basic soils texts.

Test samples are normally drained at both faces (top and bottom) and H_d is equal to half the sample height. Field drainage can occur toward the surface and to a pervious layer at depth or laterally. The difficulty in estimating the field drainage path length is one reason for lack of correlation between consolidation rate predictions and field performance; diagrams on Figure 2.31 provide guidance for estimating $H_d(\text{field})$. Even in apparently homogeneous soils the horizontal permeability coefficient (k_h) can be 2–5 times greater than the vertical coefficient (k_v) and in these cases the field consolidation times should be estimated using transformed sections to determine H_d.

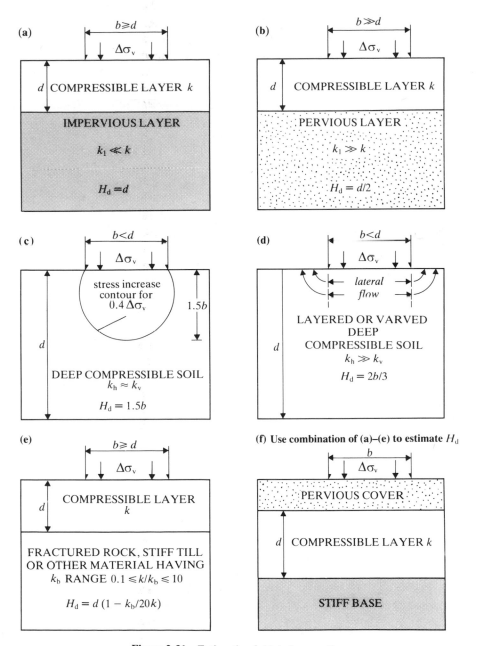

Figure 2.31 Estimating field drainage paths.

In all earth structures engineering problems where excess pore-water pressure dissipation is important to the stability of the structure, field piezometer installations to monitor this dissipation are required in support of consolidation predictions.

The classical method used in geotechnical engineering to estimate settlements is to combine elastic vertical stress distributions and compression test results. Figure 2.32, for example, gives values of the influence factors (I) at various depths (z) below a symmetrical embankment and the vertical stress increase ($\Delta\sigma_v$) at that depth is given as

$$\Delta\sigma_v = I\gamma H$$

Compressible soil layers are divided into convenient sublayers on the basis of stratigraphy, soil properties and the magnitudes of the initial stress and the stress increases, in order to calculate layer settlements to the required

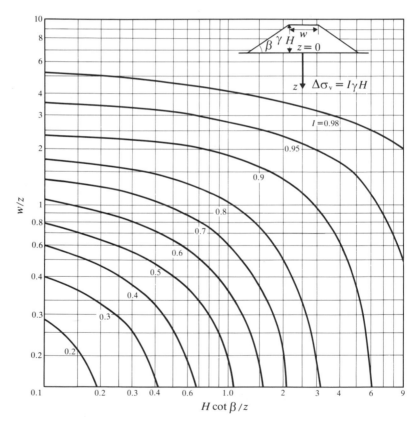

Figure 2.32 Vertical stress increase under the centerline of a symmetrical embankment (elastic solution with $\nu = 0.5$ and flexible loading).

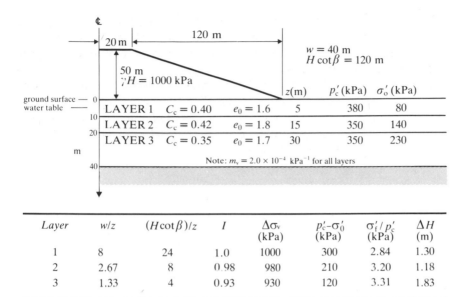

Layer	w/z	$(H\cot\beta)/z$	I	$\Delta\sigma_v$ (kPa)	$p_c'-\sigma_0'$ (kPa)	σ_f'/p_c' (kPa)	ΔH (m)
1	8	24	1.0	1000	300	2.84	1.30
2	2.67	8	0.98	980	210	3.20	1.18
3	1.33	4	0.93	930	120	3.31	1.83

Total centerline settlement by Equation 2.20 is 4.31 m

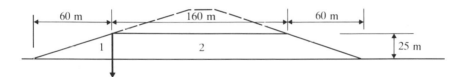

To obtain settlement under mid-slope points consider the influence of two partial embankments as shown above:

Layer	Embankment 1		Embankment 2		$\dfrac{I_1+I_2}{2}$	$\Delta\sigma_v$ (kPa)
	w/z	$(H\cot\beta)/z$	w/z	$(H\cot\beta)/z$		
1	0	12	32	12	0.98	490
2	0	4	10.67	4	0.90	450
3	0	2	5.33	2	0.82	410

Using Equation 2.20, the mid-slope settlement is calculated to be 0.87 + 0.76 + 1.16 = 2.79 m and the differential in settlement between the centerline and mid point is 1.52 m

Figure 2.33 Typical settlement calculation.

degree of accuracy. A typical calculation is outlined on Figure 2.33. The principle of superposition can be applied using elastic influence factors such that off-centreline stress increases can be estimated and influence factors can be obtained for non-symmetrical embankments, excavations (to estimate bottom heave) and other similar earthworks.

85

The major error in the classical approach to estimating settlements is that lateral movements are not taken into account. In soft soils or when the bearing capacity safety factor is low, lateral displacement of the foundation soil may result in significant additional settlements. Although empirical methods of estimating the lateral movements under embankments have been developed (Skempton & Bjerrum 1957), the trend is toward the use of modern numerical analysis techniques to predict pore-water pressures, settlements and lateral movements in earth structures engineering (see, for example, Burland 1971, Bozozuk & Leonards 1972). Piezometers and settlement gages are generally used to monitor the performance of foundation soils with pore-water pressure measurements being used to control construction rates.

3 Embankments and tunnels

Embankments can generally be constructed to great heights without concern for failure or excess settlement when founded on frictional materials or rock. Firm to stiff cohesive soils having undrained strengths in excess of 75 kPa provide sufficient bearing capacity for low embankments ($H < 15$ m) and embankment settlements will be small provided the preconsolidation pressure (p_c') is not exceeded over a significant depth in the soil profile. The exceptions to these general statements would be cases where adverse ground-water flow (artesian conditions) or ground ice (discontinuous permafrost subject to melting) could induce local failure or excessive settlements. Similarly, stiff materials have adequate strength to be self-supporting during tunnel-boring operations and sufficient stand-up time to remain stable until a tunnel liner is installed. The exceptions in the case of tunneling operations are loose or saturated frictional materials prone to raveling or piping into underground openings, and materials susceptible to squeezing or to rapid swelling and softening under stress release. These exceptions may require special construction techniques but the major problems in designing low embankments and tunnels arise when soft sediments are encountered. This chapter discusses the basic design and construction considerations related to embankments and tunnels in soft cohesive sediments.

3.1 Embankments on soft ground

A typical failure of an embankment on a clay subgrade is shown on Figure 3.1. Observations of embankment failures on soft cohesive soils indicate that both rotational failures as shown on Figure 3.2a and lateral spreading failures as shown on Figure 3.2b have developed. In either case the embankment material contributes very little to the stabilizing forces (or moments), and static bearing capacity analyses (see, for example, Wu 1976) give the theoretical safety factor for a constant undrained strength as

$$F = 5.14 Cu/\gamma H \qquad (3.1)$$

If the strength increases with depth in an ideal normally consolidated soil according to

$$Cu = z\gamma' \sin \phi'/(1 + \sin \phi')$$

87

Figure 3.1   Failure of an earth embankment by lateral spreading.

which gives $Cu = 0.33z\gamma'$ for $\phi' = 30°$, and assuming that the average strength for a failure such as those shown on Figure 3.2 occurs at a depth of about $z = 0.7H$, substitution in Equation 3.1 gives a safety factor of $F = 1.19\gamma'/\gamma$ which will always be less than unity. Fortunately, nature generally provides either a desiccated crustal zone or a sand cover over soft sediments so that the surface strength is greater than zero and a reasonable average undrained strength is provided to support embankments. Soft soils with Cu values of 20 to 40 kPa can support only modest embankments of 5 to 10 m height, however, and the design of embankments on soft ground requires considerable attention to detail.

3.1.1   Stability of embankments on soft ground

For practical purposes the factor of safety of an embankment on a soft cohesive soil has traditionally been estimated as

$$F = N_c Cu/\gamma H \tag{3.2}$$

where γH is the maximum subgrade stress due to the embankment, N_c is taken to be 5 and Cu is the average undrained strength of the soil to a depth approximately equal to the embankment height. Undrained strengths are

(a) Rotational failure

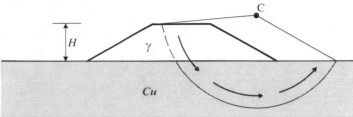

(b) Lateral spread failure

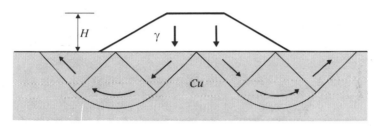

Figure 3.2 Embankment foundation failures.

normally measured by field vane testing with consolidated undrained triaxial testing to confirm the field vane results. Pilot (1972) and Bjerrum (1972) present case studies showing a correlation between the calculated factor of safety when embankments failed (using Eq. 3.1) and the plasticity characteristics of the foundation soil, and Bjerrum suggests that the vane strength should be adjusted for bearing capacity design purposes according to a graph which can be approximated as

$$Cu(\text{corrected}) = Cu(\text{vane})\left(1 - 0.5\log\frac{PI}{20}\right) \qquad (3.3)$$

At least three behavioral reasons have been proposed to account for this suggested correction:

(a) Field vane tests are carried out very quickly and plastic clays often exhibit a decrease in shearing strength with time to failure.
(b) Lateral spreading of soft plastic clay beneath an embankment loading increases the distortion and produces a work-softening effect (a decrease in shearing resistance with increased distortional strain).
(c) Undrained strength anisotropy in soft plastic soils leads to the vane measurement of strength on a vertical plane being higher than the average strength mobilized on a typical failure plane through the soil mass.

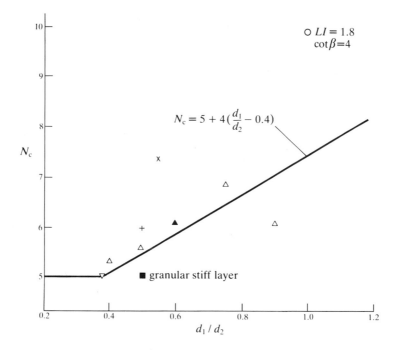

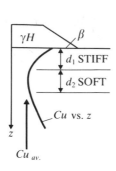

Key		
Reference	*PI*	*St*
X Wilkes (1972)	30	–
△ Pilot (1972) (four cases)	variable	8
O Ladd (1972)	15	10
▽ Dascal *et al.* (1972)	40	10
■ Eide and Holmberg (1972)	–	–
+ Raymond (1972)	40	–
▲ Lo and Stermac (1965)	21	varved

$$N_c = 5 + 4\left(\frac{d_1}{d_2} - 0.4\right)$$

O $LI = 1.8$
$\cot\beta = 4$

■ granular stiff layer

Figure 3.3 Stratigraphic effects.

All of the above plus the effects of stratigraphy (mainly the relative thicknesses of the stiffer surficial layer and the underlying soft zone, as shown on Fig. 3.3) and the effects of embankment geometry alter the application of the undrained (*Cu*) analysis to embankment stability. The problem of estimating the average undrained strength of a profile as shown on Figure 3.3 can be eliminated by using the average undrained strength of the soft zone. Using this average *Cu* the N_c value required to give $F = 1$ for a variety of embankment failures is plotted against d_1/d_2 on Figure 3.3. Thicker crustal zones increase the bearing capacity while thicker soft zones decrease the bearing capacity, but it would appear that d_1/d_2 must exceed 0.4 before any substantial increase in bearing capacity is realized. The data on Figure 3.3 indicate that the N_c value can be increased when the crust thickness exceeds 0.4 times the soft zone thickness according to the relation

$$N_c = 5 + 4\left(\frac{d_1}{d_2} - 0.4\right) \tag{3.4}$$

The effects of layering and strength anisotrophy are discussed further by Graham (1979). It is well known that lateral berms as shown on Figure 3.4 will increase the stable embankment height. The equation

$$F = \frac{N_c Cu}{\gamma(H-d)} \leq \frac{N_c Cu}{\gamma d} \tag{3.5}$$

is often used to analyze this situation. Berms have also been found to reduce the settlement of embankments by providing resistance to lateral spreading and the associated work softening or remolding of soft sensitive soils. Reduction of the embankment slope angle can achieve a similar effect, as indicated on Figure 3.4, but this is generally less efficient in terms of material quantities.

As an alternative to the undrained strength stability analysis, an effective stress approach can be adopted using effective strength parameters c' and ϕ'. In such analysis, however, the induced pore-water pressure due to embankment loadings must be estimated and this requires detailed numerical analysis combined with many simplifying assumptions in respect of the soil behavior. At the present time the confidence in safety factors obtained by this alternative appears to be less than the confidence in the traditional undrained stability analysis (Tavenas 1979), but the effective

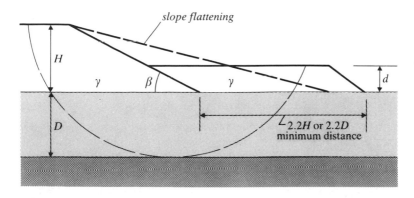

Figure 3.4 Effects of berms and slope angle.

stress approach does allow a detailed analysis of the increase in bearing capacity with time, and provides the basis for detailed analyses of lateral deformations and settlements using numerical techniques (Bozozuk & Leonards 1972).

3.1.2 Pore-water pressures under embankments

It is possible to use elastic (or modified elastic) stress analysis together with a pore-water pressure function (Eq. 2.5 or some more complex function) to estimate foundation pore-water pressures under embankments (see, for example, Burland 1971, Law & Bozozuk 1979). Incorporation of consolidation parameters also allows the dissipation of pore-water pressures during and after construction to be estimated (D'Appolonia *et al.* 1971, Leroueil *et al.* 1978) and, hence, the increase in strength with time. Measured pore-water pressures from four sites analyzed by Law and Bozozuk (1979) are plotted in a dimensionless form on Figure 3.5. These data indicate that the end-of-construction pore-water pressures under the centerline of typical embankments on deep fairly uniform soil profiles tend toward a triangular distribution, and that considerable dissipation of pore-water pressure can take place during construction. Law and Bozozuk (1979) indicate a coefficient of consolidation (c_v) of about 8×10^{-7} m^2 s^{-1} for these soils. The Kars data (Eden & Poorooshasb 1968) on Figure 3.5 show the influence of a thick weathered crust to $z = H$ with $C_v = 10^{-5}$ m s^{-1} and an underlying pervious material at $z = 2.7H$. The safety factors (by Eq. 3.2) for these embankments are between 1.4 and 1.6, however, and the pore-water pressure ratios would be expected to be much higher if the safety factor was closer to unity due to yielding of the foundation soils. In sensitive soils a decrease in shearing resistance may accompany yielding. Tavenas *et al.* (1978) indicate that the Cu/p_c' ratio for two sensitive clays of medium plasticity was found to reduce from about 0.26 before construction to about 0.22 after embankment loadings had caused yielding of the clay. This supports the proposal by Mesri (1975) that the stability of embankments on soft clays should be estimated using $Cu = 0.22\, p_c'$ in Equation 3.2.

3.1.3 Stage construction of embankments

As consolidation progresses, the void ratios in the subgrade soils are decreased and the undrained strengths increase. The preconsolidation pressure must be exceeded, however, before a major increase in undrained strength can be realized by consolidation. Figure 2.32 shows the centerline stress increases due to an embankment loading on an assumed elastic soil profile. These data are used to calculate stress increases caused by the stage construction shown on Figure 3.6. The initial embankment of height H_1 is

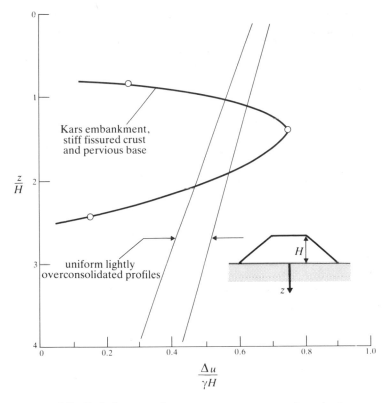

Figure 3.5 End-of-construction pore-water pressures under embankments.

designed using Equation 3.2, the minimum undrained strength and a safety factor of 1.2. The initial Cu/p'_c ratio is found to be 0.28 but the final Cu/σ'_v ratio is taken to be 0.22 which is given by Equation 2.9 with $PI = 30\%$. The effective stress in the weakest zone at 3.5 to 4 m depth at any time after construction is given as

$$\sigma'_v = \sigma'_0 + 0.82\gamma H_1 - u$$

where u would be measured using piezometers at this location. If pore-water pressure dissipation follows the pattern of Figure 2.30a and the crust is considered relatively pervious, the strength increase in the weakest zone will be fairly rapid and the weakest zone would move slightly deeper into the soil profile. After 60% average degree of consolidation ($T_v = 0.3$) the soil should achieve a fairly constant strength of about 30 kPa and the embankment height could be increased, using $F = 1.2$ in Equation 3.2, to 7 m. The final strength in the weak zone averages about 30 kPa, indicating no further increase in embankment height with further stage delay after $T_v = 0.3$ in this

93

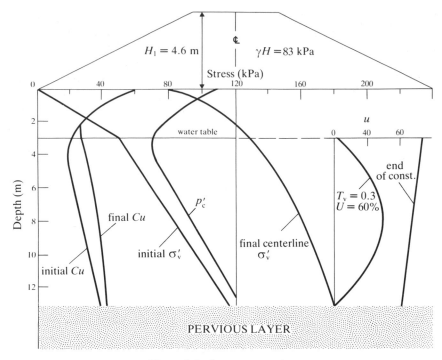

Figure 3.6 Stage construction.

case. In the case of non-sensitive soils where the ratio Cu/σ_v' might be expected to remain approximately equal to the initial Cu/p_c' ratio (given approximately by Eq. 2.7), the final embankment height would be calculated (for $F = 1.2$) as $H = 8.8$ m. In the case where the soil profile is normally consolidated, the embankment height at any time can be estimated as

$$H = \frac{N_c}{F}\left(\frac{Cu}{\sigma_v'}\right)\left(\frac{\sigma_0'}{\gamma} + \overline{U}H_1\right) \tag{3.6}$$

where $\overline{U}$ is the average degree of consolidation over the potential failure zone. It is recommended, however, that stage construction be controlled by field measurements of pore-water pressures. Stage construction of embankments on soft clays is discussed in further detail by Tavenas *et al.* (1978) who indicate that effective stress analyses can be applied (using measured groundwater pressures) to facilitate more rapid completion of stage construction. Construction of embankments on soft soils can also be facilitated by foundation drainage.

3.1.4 Foundation drainage for embankments

The rate of consolidation and hence the rate of strength increase in soft foundation materials can be increased by providing vertical sand drains to conduct water toward the ground surface where it drains laterally out through a pervious surface blanket. Additional advantages of this system are that the radial consolidation drainage toward the sand drains is usually quite rapid due to higher lateral permeabilities (particularly in varved or banded sediments) and that the sand drains provide some vertical reinforcing and reduce overall settlements. The disadvantages are the additional time delay and costs associated with the installation of the sand drains. Sand drains (usually 0.3 to 0.5 m in diameter) should be installed in sensitive cohesive soils by jetting methods because displacement methods cause extensive remolding of the soil, and hollow stem augering methods may cause local remolding and a reduction in lateral permeability. When thick sand or silt layers occur in the stratigraphy, however, overjetting is likely and a hollow stem augering method is recommended. Displacement methods are satisfactory in non-sensitive materials. The design spacings of sand drains depend on the consolidation characteristics of the soil profile and the rate of consolidation desired to facilitate the construction sequencing. Typical spacings are between 3 and 5 m in triangular patterns. Further design details are discussed by Johnston (1970) and a successful case study of an embankment constructed to about four times the short term allowable height (by Eq. 3.2) in a period of about 20 months, using a combination of lateral berms and sand drains, is reported by Ladd *et al.* (1972). According to Equation 3.4 the lateral berms account for roughly 50% of the height increase in this case study.

Relatively inexpensive wick drains (see, for example, Burke & Smucha 1981) have been used to reduce the cost of drain installation (sand drains generally cost between 3 and 5% of the embankment construction costs) and vacuum well points have been used to increase the rate of consolidation, but the advantages and confidence in the sand drain method makes this the most common method of improving soft foundation materials during embankment construction.

3.1.5 Embankments over organic soils

Very soft organic soils and peats (muskeg) usually require surface treatment prior to embankment construction. Excavation and replacement with acceptable fill is the preferred treatment, but corduroy roads have been floated over deep organic soils by placing tree trunks across the right of way to form a platform to support low embankment heights. Such roadway embankments are subject to large deflection under heavy loads, to frost action and to the formation of local sink holes.

When rockfill is available from cuttings, a displacement method is often used whereby a controlled bearing capacity failure is promoted (by blasting or mechanical ditching to sever the fibrous cover materials) to allow the rockfill to settle continuously until it is supported by stiffer subsurface materials (see, for example, Weber 1962). Most railway embankments over soft soils have been formed by displacement methods and these methods are commonly used in modern highway construction over muskeg.

With very careful control of construction, modest embankment heights can be achieved over fibrous peats without bearing capacity failures (Raymond 1969, Samson & LaRochelle 1972). The success of this type of construction depends on maintaining the integrity of the root systems in the fibrous near-surface layer, in order to provide a tensile membrane between the embankment and the underlying amorphous peat or soft organic soil. In recent years the use of geotextiles as a surface cover has enhanced the success of embankment construction over soft organic soils.

3.2 Soft-ground tunneling

Present-day soft-ground tunneling is carried out using a shield which is a cylindrical cutting ring composed of a cutting edge, a trunk and a tail section. Tunnel lining segments are usually assembled inside the tail section and the annulus around the liner is grouted with a pea gravel/bentonite grout as quickly as possible (see Fig. 3.7). Hydraulic jacks around the trunk section are used to advance the shield by pushing against the assembled liner. The shield length is typically about equal to the tunnel diameter and it is assembled in a heading at the bottom of an access shaft. Tunnel alignment is controlled by the jacking process using standard survey equipment or a light

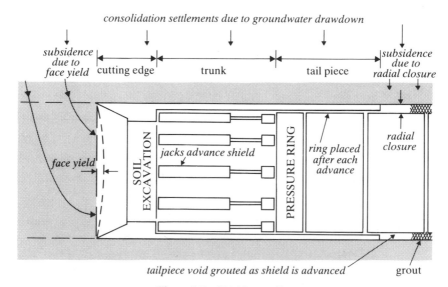

Figure 3.7 Shield tunneling.

beam to provide direction. The tunnel face can be advanced by hand excavation inside the cutting edge (benching) or by a mechanical excavator (tunnel-boring machine), with the excavated material being loaded directly or conveyed into light railcars which are moved to the shaft and hoisted for dumping. Air pressure can be used to reduce the total stress difference at the tunnel face or to reduce the hydraulic gradient in cases where high infiltration or piping could develop. Air locks and compressors are costly overhead items and there are generally strict labor regulations for working under air pressure, particularly when the air pressure is in excess of 180 kPa (see, for example, MTC 1976). For tunnels of moderate depth the cost of compressed air is approximately the same as the cost of de-watering but the air pressure will reduce surface subsidence provided that the permanent tunnel lining is sealed to maintain the groundwater level above the tunnel. De-watering may, however, be necessary to prevent collapse of saturated granular zones (Seychuck & Lahti 1979) and air pressure must be controlled to prevent large leakage of air or blowouts.

3.2.1 Tunnel stability and liners

Broms and Bennermark (1967) developed a stability criterion for exposed faces in cohesive soils which is simplified for openings that are small with respect to the depth of the opening (for moderate-to-deep shield tunneling) as

$$\frac{\gamma z}{Cu} \leq 6 \tag{3.7}$$

where γz is the total vertical stress and Cu is the undrained soil strength at the tunneling depth.

When $\gamma z > 6Cu$ difficulties have been experienced in maintaining tunnel alignment and with clay squeezing at the tunnel face. Where an air pressure (p_a) is used, the active total stress is reduced to $\gamma z - p_a$ and the stability criterion becomes

$$\frac{\gamma z - p_a}{Cu} \leq 6 \tag{3.8}$$

The numerical value of $(\gamma z - p_a)/Cu$ is often referred to as the overload factor.

Peck (1969a) supported the above stability criterion with further case records and Schofield (1981) indicates that failure developed in centrifugal model studies of tunnel faces at values of $\gamma z/Cu$ close to 5. The actual value applicable to a given tunneling project may depend on the size of the tunnel opening, the sensitivity and plasticity of the material and the rate of tunnel advance. If layers, lenses or pockets of cohesionless materials are present at the tunneling depth it may be necessary to use compressed air or pre-grouting to stabilize these materials or to de-water ahead of the tunneling operation (see Section 2.2.5).

The value of the overload factor $(\gamma z - p_a)/Cu$ will have a direct bearing on the design of a tunnel liner. For low overload factors (< 4) the material immediately surrounding the tunnel will achieve a state of pseudo-elastic equilibrium and a relatively flexible liner (such as a thin segmental concrete liner), with injection grouting of all voids between the liner and the surrounding material, would be adequate to support the compressive ring stresses. For higher overload factors it may be necessary to provide a more rigid permanent liner (reinforced concrete or bolted steel) to resist bending moments that would develop due to soil yielding and creep distortions. Considerations of stress redistribution and deformations around tunnel openings by Peck (1969a) and by Deere et al. (1970) have resulted in detailed recommendations with regard to the design of tunnel linings for ring stresses and bending moments.

In some cases tunnel liners must present a smooth interior surface and this has traditionally been achieved by erecting a corrugated steel or other temporary lining behind the shield and then casting a permanent reinforced concrete lining after the tunneling has been completed (see, for example, Eden & Bozozuk 1969). Considerable savings can be realized, however, if a simple precast segmented concrete lining can be employed. A new method of tunneling using a full-face boring machine which lends support to the tunnel face has allowed this type of unbolted liner to be placed in soft soils with high overload factors (see, for example, Palmer & Belshaw 1980).

The use of overload factors and undrained stress redistribution around tunnels to design permanent tunnel linings does not account for possible changes in effective stresses due to drawdown of the groundwater table. If a tunnel line is effectively sealed due to tight joints in segmented liners or to effective grouting of the void between the liner and the surrounding material, the groundwater table can be maintained with only a local drawdown effect as the tunnel face is advanced. If the liner is pervious after grouting, the groundwater table will be drawn down and the effective stress in the material surrounding the tunnel will increase, causing consolidation settlements. In normally consolidated or lightly overconsolidated clay soils (soft soils) the resulting consolidation will cause a further redistribution of stresses around the tunnel and may increase the bending moments induced in the tunnel lining. The maximum stress difference between the top and side of the tunnel lining would be estimated as $z\gamma \sin \phi'$ if the soil surrounding the tunnel attained a normally consolidated state. In practice this stress difference may be reduced due to arching over a relatively flexible liner. In any case the effective stress changes due to drawdown of the water table should be considered in estimating the long-term stresses on the tunnel liner. Possible influences from adjacent tunnels or other structures should also be con-

98

sidered (Peck 1969a). Stress measurements on tunnel liners in soft soils have been reported by Eden and Bozozuk (1969) and by Palmer and Belshaw (1980).

3.2.2 Ground subsidence due to tunneling

Settlements above tunnels in soft ground result from three basic factors as indicated on Figure 3.7.

(a) Undrained deformations producing ground loss at the tunnel face as the shield is advanced: these will be increased by high overload factors and by poor control of jacking and will be minimized by boring machines that provide full-face support and a more rapid advance of the face.
(b) Undrained deformations produced by closure of the soil around the tunnel liner: these will be increased by high overload factors and will be minimized by good quality control on grouting of the annular space behind the tailpiece as the shield is advanced.
(c) Drained deformations produced by drawdown of the groundwater table (consolidation settlements): these will be increased by high overload factors (since the effective stresses will exceed p'_c with minimal drawdown) and will be decreased by effective grouting to decrease the permeability of the lining.

It is apparent that the ground subsidence is related to the overload factor for any given method of tunneling and overload factors less than 4 are recommended in order to minimize subsidence. From case studies summarized by Peck (1969a) and Schmidt (1974) the volume of ground loss (ΔV) in single shield tunnels with good construction control appears to be related to the volume of the tunnel excavation (V) as

$$\frac{\Delta V}{V} \simeq \left(\frac{Cu}{E_u}\right) \exp \frac{\gamma z - p_a}{2Cu} \tag{3.9}$$

and the undrained modulus (E_u) for soft soils is generally about 200 to 700 times Cu. For very sensitive soils or for poor construction control (e.g. excavating ahead of the shield, poor jacking techniques) the ground loss estimate should be increased by a factor of about 3. If the settlement trough is approximated by a normal probability curve (Peck 1969a) with maximum settlement δ_m and inflection points at distances i on each side of the tunnel centerline, the location of the inflection points may be approximated as

$$i = 0.2(d + z) \tag{3.10}$$

99

where d is the tunnel diameter and z is the tunneling depth. The settlement at the inflection point locations is $0.6\delta_m$ and δ_m can be approximated as

$$\delta_m = \frac{2\Delta V}{d+z} \tag{3.11}$$

For twin tunnels the value of d in Equations 3.10 and 3.11 should be replaced by $(d + x)$ where x is the distance between the tunnels. Peck *et al.* (1969) provide additional data for refinement of the above approximations.

It is of interest to compare estimated ground loss to radial closure around the liner. Assuming that a 6 m diameter tunnel excavation would leave a 40 mm annular space between the outer diameter of the shield and the outer diameter of the liner, the closure of this void would represent a value of $\Delta V = 0.02V$. Using a typical value of 400 for the ratio of the undrained modulus to the strength and an overload factor of 5 in Equation 3.9, the estimated value of ΔV is $0.03V$. This comparison indicates that roughly 70% of the ground settlement might be due to closure of the annulus behind the shield, and the general belief that it is important to grout this void space as quickly as possible as the shield is advanced is supported.

The largest subsidence due to tunneling in soft compressible ground would occur if the ground water is permanently lowered significantly. From Equation 2.18 drawdown to the tunneling depth in a normally consolidated soil would produce a maximum surface settlement, for tunnels of relatively small diameter, of

$$\Delta H = \frac{hC_c}{1 + e_0} \log\left(1 + \frac{\tfrac{1}{2}h\gamma_w}{z_1\gamma + \tfrac{1}{2}h\gamma'}\right) + \frac{z_2 C_c}{1 + e_0} \log\left(1 + \frac{h\gamma_w}{z_1\gamma + z_3\gamma'}\right) \tag{3.12}$$

where h is the depth below the original groundwater table to the mid-depth of the tunnel, z_1 is the soil cover above the water table, z_2 is the depth below the tunnel mid-depth to a stiff base and $z_3 = h + d + 0.5z_2$. Using $z_1 = 0$, $C_c = 0.4$, $e_0 = 1.5$, $\gamma = 1.6\gamma_w$ and $h = z_2 = 10$ m, for example, the maximum subsidence due to full drawdown would be estimated from Equation 3.12 to be about 0.7 m. This subsidence trough in a uniform soil profile would have a shape similar to the two-dimensional drawdown curve and would extend laterally to distances in excess of $5h$. Both the maximum subsidence and the extent of subsidence would be functions of time and the rate would be governed by the coefficient of consolidation (C_v) and the tunneling depth below the water table (h). Permanent lowering of the groundwater table in the above case would produce a subsidence volume of about 35 m^3 per meter of strike length, which is nearly twice the unit volume of a 5 m diameter tunnel. It is apparent that the tunnel lining and void grout combination must have a relatively low permeability in order to reduce subsidence when tunneling at significant depths below the water table.

Two case studies serve to emphasize subsidence problems. Attewell (1977) reports a case study where a 4.25 m diameter tunnel was shield-driven and hand-excavated at a depth of 10 m below the water table in a soft normally consolidated alluvial silty clay using air pressure to reduce the overload factor to 5.0. Short-term subsidence (δ_m) was about 30 mm as expected but the long-term subsidence ($\Delta \dot{H}$) reached 140 mm after 350 days despite the use of grout in the void behind the shield and periodic back-grouting at 690 kPa pressure to try to seal the tunnel from infiltration. Palmer and Belshaw (1980) report a case study where a 2.16 m tunnel was shield-driven using a full-face boring machine at a depth of 10 m below the water table in a soft sensitive silty clay without air pressure and with an overload factor of about 5. The short-term subsidence (δ_m) was about 45 mm after 20 days and the additional long-term settlement was less than 20 mm during the first year after construction. A bentonite grout was injected into the tailpiece void during shield advance, and piezometer data showed little drawdown. From flow considerations the authors indicate that a zone surrounding the tunnel had a hydraulic conductivity about 0.05 times the average soil permeability, and that this relatively impervious zone effectively sealed the tunnel. The relatively high short-term ground surface subsidence in this case is attributed to the sensitivity of the soil (local remolding and consolidation around the tunnel) while the relatively high long-term subsidence in the case reported by Attewell was due to consolidation of the material surrounding the tunnel. Back-grouting in that case was ineffective in sealing the tunnel possibly because the grout pushed out into the soft materials instead of sealing seepage channels around the tunnel lining. To be effective in sealing the tunnel against high seepage infiltration, the grout should be injected into the tailpiece void before this void has time to close around the liner.

3.3 Problems on bearing capacity and tunnels

3.3.1 Example problems

Example problem 1 Estimate the critical height of an embankment having $\cot \beta = 2$ and $\gamma = 18$ kN m^{-3} constructed rapidly on the soil profile shown on Figure 3.6 (assume $PI = 30$). Compare this with the estimated ultimate bearing capacity of a 2 m width strip footing placed on the surface of this soil and explain the reason for any difference in bearing capacity.

SOLUTION The average Cu over the soft zone is taken to be 22 kPa and $d_1/d_2 = 0.7$. From Figure 3.3 an N_s value of 6 could be assumed. Then from Equation 3.2, $H = 7.3$ m at $F = 1$. The bearing capacity is $H\gamma$ or 132 kPa.

For the strip footing three separate estimates could be made as follows:

101

(a) Using $Cu_{avg} = 38$ kPa over the top 3 m the ultimate bearing capacity is q $= 5Cu = 190$ kPa.
(b) Using the analysis of Mitchell *et al.* (1972a) the ultimate bearing capacity is 200 kPa.
(c) Using an influence factor of $0.6q$ for the bottom of the crust (see standard soil mechanics texts) and limiting the vertical stress increase at this level to $\Delta\sigma_v = 5(22) = 110$ kPa, gives the ultimate bearing capacity as $(110/0.6) = 183$ kPa.

Of course the strip footing would have a factor of safety in the order of two or greater to limit settlements. The reason that the embankment bearing capacity is lower is that the embankment influence factor at the soft soil depth is close to unity. The overlying stiff layer and the increased stiffness below the soft layer act as reinforcement, however, and allow an increase in the bearing capacity factor N_c.

Example problem 2 Compare predicted centerline subsidence with that reported by Attewell (1977) as noted in Section 3.2.2. Use $C_c = 0.30$, $E_u = 600Cu$, $\gamma = 16$ kN m^{-2}, $e_0 = 1.8$. The depth to the stiff base below the tunnel is 3 m and the depth of material above the water table is 5 m.

SOLUTION Using Equation 3.9 the immediate ground loss is estimated as $\Delta V/V = 0.02$. From Equation 3.11 the maximum immediate subsidence is given as $\delta_m = 0.029$ m. From Equation 3.12, the long-term centerline settlement due to full drawdown of the water table is given as 0.26 m. The prediction of immediate settlement agrees favorably with the 30 mm reported but the reported long-term settlement of 140 mm is only 54% of the predicted value. It might be concluded that the water table was not fully drawn down to the tunnel level but it is also possible that the volumes of grout periodically injected into the soil to try to seal the tunnel accounts for the difference between the calculated and the observed long-term subsidence.

3.3.2 Assignment problems

Problem 1 An embankment of 15 m height is to be constructed on the soil profile shown on Figure 3.6 within a period of 18 months. Assume $PI = 30$, $C_v = 1 \times 10^{-7}$ m^2 s^{-1} in the crustal zone and 1×10^{-6} m^2 s^{-1} in the underlying clay, $k_h = 2k_v$, $St = 8$ for the foundation soil and $\gamma = 18$ kN m^{-3} for the embankment.

(a) Consider the possibilities of constructing this embankment by stage construction with sand drains and with berms.
(b) If compacted fill for berms cost $2.80 per m^3 plus a cost of $200.00 per meter of embankment strike length for additional right of way, and sand drains cost $4.25 per meter of sand drain, recommend the most economical method of construction.

Problem 2   It is proposed to construct a 4 m diameter gravity sewer tunnel in the soil profile shown on Figure 3.6. The soil unit weight is $17 \, \text{kN m}^{-3}$. The shield-driven tunnel could be at a depth of 8 m or at a depth of 12 m but the deeper tunnel would be preferable to the design of the pumping facilities at the sewage treatment plant. Use $Cu = 0.002 \, E_\text{u}$.

(a) Estimate the form and magnitude of the ground surface settlements to be expected in each case if it is assumed that the groundwater table will be maintained.

(b) Estimate the form and magnitude of the ground surface settlements in each case if the water table is drawn down to the sewer level.

(c) Discuss any additional problems to be anticipated if the tunnel is driven at 12 m depth.

Problem 3   In a very soft normally consolidated submarine soil the strength is likely to be as low as $Cu = 0.18z\gamma'$. Calculate the air pressure requirement for shield tunneling at depths between 10 and 30 m if $\gamma = 16.8 \, \text{kN m}^{-2}$. Comment on the feasibility of shield tunneling in such soft sediments.

4 Slope stability

Earth movement is the general term applied to all vertical or lateral movements of portions of the surface of the Earth. Of major engineering significance are slope movements and ground subsidence. This chapter is concerned with slope movements.

The most violent earth movements occur as stress-relief shear slips on major fault systems associated with tectonic plate boundaries and can involve several meters of almost instantaneous lateral movement producing a violent earthquake. The San Francisco earthquake of 1906 occurred due to slip on the San Andreas Fault and is a classic example of this type of earth movement. The shocks due to such rapid movement of large masses of earth can induce ground heaving, soil liquefaction and landslides. Slope movements can also be initiated by combinations of natural oversteepening of slopes, unusually high groundwater pressures and progressive weakening of the slope material The huge rockslide that obliterated the village of Frank, Alberta, in 1903, killing 66 people (McConnell & Brock 1904), and the rapid earth flow that buried the village of Notre-Dame de la Salette, Quebec, in 1908, killing 33 people (Ells 1908), are unfortunate examples of slope movements. Although these particular examples are three-quarters of a century old and substantial progress has been made in delineating the factors contributing to geologically hazardous terrain, recent catastrophic slope failures have occurred (see, for example, Tavenas *et al.* 1971, Dishaw 1967). Route selection and resources development planning activities should always consider natural slope stability and the effects of construction or extraction of materials on the stability.

On a less catastrophic scale, innumerable translatory or rotational slides, slumps, block slides and debris topples occur each year due to erosion and weathering in actively developing landscapes. These are very costly in terms of property loss and maintenance to existing facilities. On many residual soil slopes, very slow continuous downslope movements create a major maintenance problem for ground transport systems. Urban development along young river valleys is also a factor contributing to the property losses due to landslides. Slope movements are a natural part of the geomorphological development of young landscapes and this should be considered in the planning and approval stages of developments.

4.1 Types of slope movements

Slope movements have been classified with regard to morphology and material type by various researchers and one of the most detailed classifica-

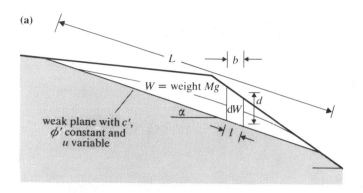

Figure 4.1 Translatory landsliding. (a) Schematic diagram. (b) Translatory landslide in an active clay till believed to be triggered by removal of vegetation and placement of road fill. This type of landsliding is associated with weak planes and high groundwater pressures and it usually develops as seasonal downslope movement but can accelerate into a catastrophic slide. (Photo by C. MacKay.)

tions has been published by Varnes (1978). Some excellent oblique sketches and photographs of various types of landslides can also be found in a publication by Royster (1980). Stereo air photos of practically all types of slope movements are contained in a manual by Mollard (1973).

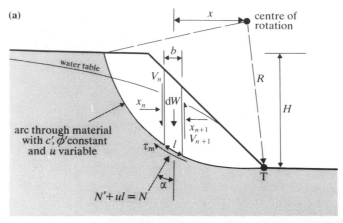

(a)

(b)

Figure 4.2 Rotational landsliding. (a) Schematic diagram. (b) Small rotational landslide in a fissured clay believed to have been triggered by a back-up of ground water due to freezing of the seepage surface. Small slides are often triggered by toe erosion and rain or snow infiltration to create high groundwater pressures. Rotational sliding is common in homogeneous materials and occurs at a rapid rate. (Photo courtesy of the Division of Building Research, NRC, Canada.)

For the purpose of engineering analysis, slope failures are classified on Figures 4.1 through 4.5 according to the mechanisms of failure. Major causes, material types and relative speed of movement are also noted on these figures. Simple caving mechanisms such as cliff falls (due to undercutting), toppling in a columnar rock (due to weathering or ice action) and the analysis of avalanches (rock, snow or debris) in mountainous terrain are not included in the analyses presented in this text. Photo examples of various slope failures appear in Chapter 1 and further examples are contained in this chapter.

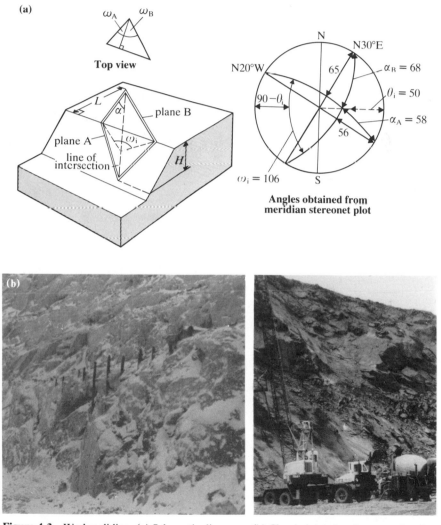

Figure 4.3 Wedge sliding. (a) Schematic diagrams. (b) Closely jointed rock mass and wedge type failure. When planes of weakness intersect and daylight in a slope face following excavation, a rapid failure may result. Joint weathering may reduce the shearing resistance in the planes of weakness and result in delayed failure. Earthquakes also trigger this type of sliding by reducing the interlocking on rough intersecting planes of weakness. (Photos courtesy of John D. Smith Engineering Associates Ltd.)

107

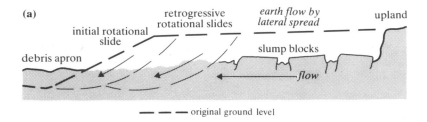

(a)

initial rotational slide

retrogressive rotational slides

earth flow by lateral spread

upland

debris apron

slump blocks

flow

— — — original ground level

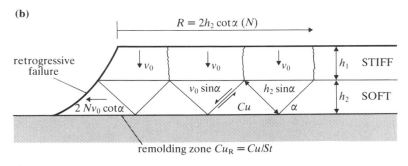

(b)

$$R = 2h_2 \cot\alpha\ (N)$$

retrogressive failure

v_0 v_0 v_0

h_1 STIFF

$v_0 \sin\alpha$ $h_2 \sin\alpha$

$2\,Nv_0 \cot\alpha$ Cu α

h_2 SOFT

remolding zone $Cu_R = Cu/St$

(c)

Figure 4.4 Rapid earthflow by lateral spread. (a) Schematic diagram of rapid earthflow. (b) Schematic diagram of Odenstad failure mechanism. (c) Rapid earthflow by retrogressive rotational sliding and lateral spreading of soft clay overlain by stiff silt and sand deposits. This type of landsliding is triggered by an initial bank failure when the upland area is composed of sensitive materials susceptible to earthflows. The very rapid and destructive flow of earth materials is similar in appearance to liquefaction failures and weak layers or layers susceptible to liquefaction can be instrumental in promoting earthflows. (Photo courtesy of the Division of Building Research, NRC, Canada.)

4.2 Slope stability analyses

Upper-bound plastic limit analysis is the basic tool for estimating the safety factor, F, of a slope. The basic approach is quite simple:

(1) Assume a kinematically admissible velocity field or failure mechanism.
(2) Calculate

$$F = \frac{\text{energy dissipated in the movement}}{\text{energy causing movement}}$$

The method allows fairly complex geometric and stratigraphic conditions to be analyzed but requires some engineering judgment in its more general applications. Figure 4.6 presents an idealized situation to which the general principles can be applied as follows:

(a) Movement of slope DOTS at velocity v_0 is proposed to develop by sliding on a weak plane AS and rotational movement in a zone OBA. Sliding along plane DB completes the mechanism.

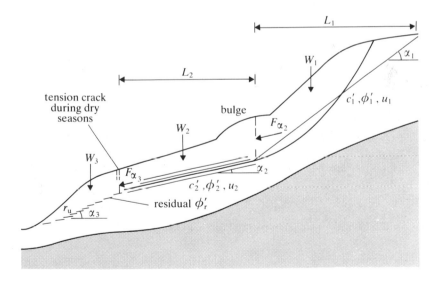

Figure 4.5 Slow earthflow or debris flow. This type of sliding is associated with right-of-way maintenance problems in mountainous terrain. Air photo observations may be valuable in assessing the relative stability of various sections of the sliding mass. Movements usually increase during rainy seasons and may stop during dry seasons. Rates of movement can be in the order of several meters per year. (For photo see Fig. 1.15.)

109

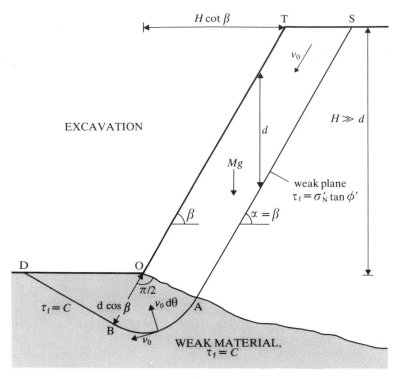

Figure 4.6 Idealized potential failure mechanism.

(b) Neglecting second-order terms, the time rate of energy release is given as

$$E_i = Mg \sin \alpha \, v_0 = \gamma dH \cot \alpha \, v_0 \sin \alpha$$

(c) Neglecting second-order terms the time rate of energy dissipation is given by the following:

$$E_d \text{ (on plane AS)} = \gamma dH \cot \alpha \cos \alpha \tan \phi' \, v_0$$

$$E_d \text{ (on arc BA)} = \int_{\theta=0}^{\pi/2} C v_0 R \, d\theta = C v_0 d \cos \alpha \, (\pi/2)$$

$$E_d \text{ (in radial shear zone OAB)} = \int_{\theta=0}^{\pi/2} C d \cos \alpha \, v_0 \, d\theta = C d \cos \alpha \, v_0 \, (\pi/2)$$

$$E_d \text{ (on plane BD)} = C d \sin \alpha \, v_0$$

(d) Then $F = \dfrac{\gamma dH \cot \alpha \cos \alpha \tan \phi' \, v_0 + \pi v_0 C d \cos \alpha + C d \, v_0 \sin \alpha}{\gamma dH \cot \alpha \, v_0 \sin \alpha}$

$$= \frac{\tan \phi'}{\tan \alpha} + \frac{C}{\gamma H} \left(\pi + \frac{1}{\cot \alpha} \right) \tag{4.1}$$

110

This analysis indicates that when $\alpha > \phi'$ the stability of the slope depends on the weak material maintaining a constant shearing resistance, C. This is equivalent to assuming that the material behaves in a perfectly plastic manner such that the shearing resistance will not decrease if small movements or strains do develop. Also in this case the engineer must be concerned about possible long-term strength decreases due to weathering and stress release in the plane of weakness and in the exposed weak material. In such cases the earth structures engineer should subject core samples to accelerated weathering (wetting–drying and, if applicable, freezing–thawing cycles) followed by testing to determine the effect of weathering on the stress–strain properties of the materials.

Some important facts about upper-bound analyses should be noted:

(a) These analyses assume perfectly plastic stress–strain characteristics.
(b) The safety factor can be expressed as

$$F = f_1 (\tan \phi') + f_2(c'/\gamma H) \qquad (4.2)$$

where f_1 and f_2 are functions of the slope geometry and $c'/\gamma H$ and $\tan \phi'$ are dimensionless parameters.
(c) Equation 4.1 may not express the lowest safety factor for the slope since another mechanism of failure may develop at a lower energy level (a chain is only as strong as its weakest link), but if $F < 1$ the slope will fail (presuming that the operative strength parameters are correct).

The above facts are true of any upper-bound slope stability analysis and it is the job of the geotechnical engineer to find the weakest link (lowest energy failure). It should also be noted that additional energy from vibrations (particularly from blasting or earthquakes) and from groundwater seepage might also be active in promoting failure. Due to published *post-mortem* analyses of a large number of various types of slope failures, however, confidence in the commonly assumed mechanisms and safety factor calculations has been established.

4.2.1 Translatory landsliding analysis

The mechanism of failure shown on Figure 4.1 involves rigid body translation with no internal dissipation of energy in the soil mass. The safety factor can then be expressed as

$$F = \frac{\text{resisting forces}}{\text{driving forces}}$$

111

Considering an isolated slice of the sliding mass of width b, as shown on Figure 4.1, the safety factor of that slice is given as

$$F_s = \frac{c'l + (dW \cos \alpha - ul) \tan \phi'}{dW \sin \alpha}$$

where dW is the weight (Mg) of the slice and u is the pore-water pressure at the base of the slice. Since $dW = \gamma bd$, $l = b/\cos \alpha$ and defining a pore-water pressure parameter $r_u = u/\gamma d$,

$$F_s = \frac{c'l}{dW \sin \alpha} + \frac{\tan \phi'}{\tan \alpha}(1 - r_u \sec^2\alpha)$$

Although the factors of safety of various slices will differ slightly and this will generate interslice forces, the interslice forces must sum to zero and will be small whenever α is constant. It is sufficiently accurate to express the overall safety factor as

$$F = \frac{c'L}{W \sin \alpha} + \frac{\tan \phi'}{\tan \alpha}(1 - r_u \sec^2\alpha) \tag{4.3}$$

where an average value of r_u is obtained by averaging the r_u values for equal-width slices covering the entire failure plane.

When the angle of the weak plane (α) approaches the slope angle (β) the mechanism is referred to as infinite slope sliding because the safety factor is independent of the slope height and is given as

$$F = \frac{\tan \phi'}{\tan \alpha}(1 - r_u \sec^2\alpha) + \frac{c'}{\gamma d}\left(\frac{2}{\sin 2\alpha}\right) \tag{4.4}$$

In dry loose sand the stable angle will then be $\alpha = \phi'$, and is called the angle of repose. Where strongly cemented materials stand at slopes in excess of $45°$, tension cracks may form near the crest and may fill with water. In such cases an additional hydrostatic driving force should be considered.

4.2.2 Rotational landsliding analysis

The mechanism of failure shown on Figure 4.2 involves rigid body rotation with no internal dissipation of energy, and the factor of safety can be written as

$$F = \frac{\text{resisting moments}}{\text{driving moments}}$$

112

Early solutions to this mechanism disregarded the interslice forces shown on Figure 4.2b. That these forces, particularly the horizontal interslice forces, are significant can be demonstrated by the simple fact that a slice located directly below the center of rotation (C) has an independent safety factor of infinity since $\alpha = 0$ at that point (i.e. driving moments equal zero). Indeed, the upper part of the rotational mass provides most of the driving moments while the lower part provides most of the resisting moments. In order to improve earlier analyses, Bishop (1955) developed a solution that incorporates horizontal interslice forces by virtue of the fact that the analysis ensures that the safety factors of all slices are identical. This was done by defining the mobilized shearing resistance as

$$\tau_m = \frac{c'}{F} + \sigma'_N \frac{\tan \phi'}{F} = \frac{c'}{F} + \left(\frac{N}{l} - u\right) \frac{\tan \phi'}{F}$$

and the safety factor as

$$F = \frac{\Sigma \tau_f . l . R}{\Sigma dW . x} = \frac{\Sigma c'l + (N - ul) \tan \phi'}{\Sigma \, dW \sin \alpha}$$

since $x = R \sin \alpha$.

The vertical total force is given as

$$N \cos \alpha = dW - \tau_m l \sin \alpha - (V_{n+1} - V_n)$$

If it is assumed that $(V_{n+1} - V_n)$ is negligible in comparison with the other forces, then

$$N \cos \alpha = dW - \frac{c'l \sin \alpha}{F} - N \sin \alpha \frac{\tan \phi'}{F} + ul \sin \alpha \tan \phi'/F$$

and

$$N = \frac{dW - \dfrac{c'l \sin \alpha}{F} + ul \sin \alpha \tan \phi'/F}{\cos \alpha + \sin \alpha \tan \phi'/F}$$

or

$$N - ul = \frac{dW - \dfrac{c'l \sin \alpha}{F} - ul \cos \alpha}{\cos \alpha + \sin \alpha \tan \phi'/F}$$

113

Substituting this equation into the equation for F above gives

$$F = \frac{1}{\Sigma\, dW \sin \alpha} \sum \left(c'l + \frac{(dW - \dfrac{c'l \sin \alpha}{F} - ul \cos \alpha) \tan \phi'}{\cos \alpha + \sin \alpha \tan \phi'/F} \right)$$

$$= \frac{1}{\Sigma\, dW \sin \alpha} \sum \left(\frac{[c'l \cos \alpha + (dW - ul \cos \alpha) \tan \phi'] \sec \alpha}{1 + \tan \alpha \tan \phi'/F} \right)$$

Since $l \cos \alpha = b$, this equation can be written as

$$F = \frac{1}{\Sigma\, dW \sin \alpha} \sum \left(\frac{[c'b + (dW - ub) \tan \phi'] \sec \alpha}{1 + |\tan \alpha| \tan \phi'/F} \right) \tag{4.5}$$

Equation 4.5 is developed following Bishop (1955) and is known as the Bishop simplified solution. The development of this equation is simplified by using two definitions of safety factor and these are theoretically the same only at $F = 1$. The loss in theoretical accuracy as F increases above unit is not sufficient, however, to be of concern in the application of Equation 4.5 to practical slope stability problems. It should be noted that the original Bishop simplified solution does not use the modulus of $\tan \alpha$ as used in the denominator of Equation 4.5. This modification is included in order that $\tan \alpha$ is not taken as a negative number for slices to the left of the center of rotation (see, for example, Mehta 1973). Although more rigorous rotational landsliding analyses are available, as well as analyses of non-circular failure surfaces, analyses that incorporate anisotropic strength parameters and analyses that account for the fact that the circular failure is not strictly two-dimensional, these research efforts have generally indicated that such analytical improvements produce only small percentage effects in calculated safety factors (Bishop & Morgenstern 1960, Janbu 1973, Fredlund 1978). The simplified Bishop analysis is used in practice because it has been shown to give reliable results in a large number of case records. Morgenstern and Sangrey (1978) provide further discussion of the applicability of various slope stability solutions.

The factor of safety calculation for the rotational landsliding analysis is subject to a choice of the center of rotation and one point on the failure surface. The critical arc is defined as that circular arc which gives the lowest value for F. The critical arc is found by calculating F for a number of trial centers of rotation – a digital computer is normally used in this search and safety factors for selected centers can be contoured within an assigned search area. Slope stability computer programs are generally available – most programs calculate F for about 100 trial centers in an assigned search area using about 50 equal-width slices for each trial arc and iterating to ± 0.005 in the value of F. To define a trial arc, the trial center and one other

114

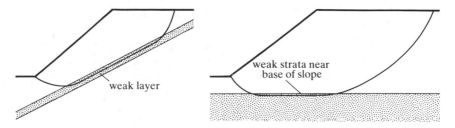

Figure 4.7 Examples of non-circular failure surfaces.

point must be specified; when $\tan \phi' > 0.1$ or the strength increases with depth such that $\Delta c/\Delta z > 0.1\gamma$, the critical arc will pass through the toe of the slope (Terzaghi 1943, Gibson & Morgenstern 1962). In other cases, and as a check on the increase of F with depth of failure, deeper points (below the toe elevation) are selected for trial circles. The following points should be noted for computer solutions:

(a) They can accurately represent the two-dimensional geometry of the earth mass including stratigraphic variations, erosional effects, crest loadings, benching, submergence, reservoir drawdown, dynamic loadings and other factors.

(b) They can accurately represent the groundwater pressures at all points in the slope using flow nets or nodal point tabulations such that approximating an average r_u for the slope is avoided.

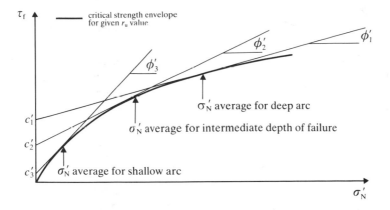

Figure 4.8 Analysis of field failures. Critical stress analysis: slope geometry, ground water (r_u) and a range of ϕ' values input to computer. Values of c' that give $F = 1$ for each ϕ' value and average τ_f, σ'_N are calculated. Laboratory strength data can be compared directly to the critical strength envelope and the location of the critical arc (found by probing or vane testing) can be compared to the critical arc computed for each strength combination. Groundwater pressures can also be compared with the analytical results.

115

(c) They can accurately represent strength variations due to anisotropy, stratigraphy, fissuring or softening. Often programs which specify non-circular failure surfaces are used where there are layers, planes or zones of weakness (Morgenstern & Price 1965, Krahn *et al.* 1971) such as shown on Figure 4.7.

(d) They can analyze field failures (*post-mortem* analysis) to determine combinations of c' and ϕ' which would give $F = 1$ and thus establish a field critical strength envelope as shown on Figure 4.8 (Kenney 1967b).

(e) They can analyze combined embankment slope stability and foundation failures using effective stresses for stage construction purposes or as an alternative to the standard bearing capacity analyses (see, for example, Raymond 1973).

If a material possesses a constant strength ($\phi' = 0$), the Bishop simplified equation (Eq. 4.5) reduces as follows:

$$F = \frac{\Sigma\, c'b \sec \alpha}{\Sigma\, dW \sin \alpha} = \frac{\Sigma\, c'l}{\Sigma\, dW.x/R} = \frac{c'LR}{\Sigma\, dW.x} = N_s \frac{c'}{\gamma H} \qquad (4.6)$$

and it is found that deeper failure arcs become critical. In this case the extent of the failure is limited by a stiff base and an additional dimensionless parameter $D = d/H$ (where d is the depth to a stiff base) is needed to characterize the failure (Taylor 1937).

If a material is cohesionless ($c' = 0$), the Bishop simplified equation reduces to

$$F = \tan \phi' \, \Sigma \left(\left(1 - \frac{ub}{dW} \sec^2\alpha \right) \Big/ \tan \alpha \right)$$

and computer analyses have shown that very shallow failure arcs are critical. The physical reason for this is that the normal stresses increase on deeper circles giving an increased frictional resistance. A very shallow failure arc will approach the straight line failure analyzed in Section 4.2.1 where α is constant. Since $r_u = ub/dW$ the Bishop equation then becomes

$$F = \frac{\tan \phi'}{\tan \alpha} (1 - r_u \sec^2\alpha)$$

which is identical to Equation 4.4 with $c' = 0$.

Equation 4.5 can be rewritten using $r_u = u/d\gamma$ and $dW = bd\gamma$ as

$$F = \frac{1}{\Sigma\, dW \sin \alpha} \sum \left(\frac{[c'b + dW(1 - r_u) \tan \phi'] \sec \alpha}{1 + |\tan \alpha| \tan \phi'/F} \right)$$

which has the form

$$F = m - n(r_u) \qquad (4.7)$$

where m and n are constant for a given dimensionless slope defined by $c'/\gamma H$ = constant, $\tan \beta$ = constant and $\tan \phi'$ = constant (Bishop & Morgenstern 1960). Values of m and n can be obtained from computer analysis of rotational sliding on a critical arc and a method for obtaining an average r_u value is outlined in Section 4.3.1.

4.2.3   Wedge landsliding analysis

Wedge failures are common in jointed rocks with steeply dipping intersecting planar discontinuities. Surveys of joint sets, bedding planes and faults in the rock can be plotted on a meridian stereonet, as indicated on Figure 4.3, to determine whether a wedge failure can occur. The major planes A and B on Figure 4.3 can be established from pole contouring after a detailed joint survey has been carried out (see, for example, Hoek & Bray 1974). Analysis of the wedge failure mechanism is presented below.

By resolving the normal and downdip components of gravity forces on the sliding block, the normal effective forces on planes A and B are obtained as

$$N'_A = W (\sin \theta_A \cos \alpha_A / \sin \omega_i) - u_A A_A$$

$$N'_B = W (\cos \theta_A + \sin \theta_A \cos \alpha_A / \tan \omega_i) - u_B A_B$$

where W is the sliding block weight

$$W = Mg = \frac{\gamma H L^2}{6} (\tan \omega_A + \tan \omega_B)$$

ω_A is the horizontal angle subtended by the strike of plane A and the line of intersection, ω_B the horizontal angle subtended by the strike of plane B and the line of intersection, θ_A the dip of plane A, α_A the angle measured in plane A between the strike of plane A and the line of intersection, ω_i the angle subtended by the two planes measured perpendicular to the line of intersection, u_A and u_B are the average water pressures in planes A and B respectively, A_A the area of plane A

$$A_A = \frac{HL}{2 \sin \theta_i} \frac{\sin \alpha_B}{\cos \omega_B}$$

θ_i is the dip of the line of intersection, and α_B the angle measured in plane B between the strike of plane B and the line of intersection.

117

The driving force along the line of intersection is given by

$$F_D = W \sin \theta_A \sin \alpha_A$$

and the theoretical safety factor is then

$$F = \frac{c'_A A_A + c'_B A_B}{F_D} + \frac{N'_A \tan \phi'_A + N'_B \tan \phi'_B}{F_D} \qquad (4.8)$$

It is commonly found that discontinuities under low normal effective stresses exhibit only frictional shearing resistance (i.e. $c'_A = c'_B = 0$) and the assumption that $c' = 0$ is generally used in analyzing wedge failures in rock (Jaeger 1971). Hoek and Bray (1974) present design charts for wedge failures in dry cohesionless rocks. By approximating the groundwater condition by an average r_u value, Equation 4.8 can be simplified for the $c' = 0$ condition to estimate the safety factor as

$$F = (1 - r_u) \left[\frac{\cot \alpha_A \tan \phi'_A}{\sin \omega_i} + \left(\frac{\cot \theta_A}{\sin \alpha_A} + \frac{\cot \alpha_A}{\tan \omega_i} \right) \tan \phi'_B \right] \qquad (4.9)$$

Wedge type failures may also develop in drumlinized till and fissured clay slopes (Skempton *et al.* 1969, Sauer 1978, McGown & Radwan 1976). Site investigation and testing procedures should be designed to investigate discontinuities in these types of materials.

4.2.4 Earthflow analysis

Rapid earthflow as depicted on Figure 4.4a is initiated by rotational failure in soils of a loose saturated or sensitive structure such that the gravitational energy released in the sliding process is sufficient to move most of the landslide debris out of the crater. Retrogressive bank failure can then develop as shown on Figure 4.4a. Such retrogression can theoretically be analyzed using Equation 4.5 but the pore-water pressure distribution would be modified by the short-term strains that develop within the slope due to the stress changes initiated by the first rotational slide. Mitchell and Markell (1974) conclude from both theoretical considerations and field observations that landsliding will terminate after three or four retrogressive rotational failures unless the undrained strength of the soil (Cu) is exceeded on a critical failure surface. They define such minor retrogressions as retrogressive flowslides and indicate that these occur in fairly homogeneous sensitive soils where $(\gamma H / Cu) < 6$. When $\gamma H / Cu > 6$ within the slope profile, particularly where soft sensitive materials are overlain by a saturated weathered crust or by saturated cohesionless soils, large rapid earthflows are a potential form of landsliding. The rapid earthflow potential can be analyzed using a mechanism

118

suggested by Odenstad (1951) and shown on Figure 4.4b. The gravitational energy causing the mechanism to develop is given as

$$E_i = \gamma h_1 R v_0$$

whereas the energy dissipation with time is calculated as

$$E_d = \frac{Cu R v_0}{\sin^2\alpha \cot \alpha} + \frac{Cu_R R^2 v_0}{2h_2}$$

where it is assumed that the sliding base is quickly remolded by the large displacements required to develop the mechanism. The various components of velocity are shown on Figure 4.4b and it is assumed that $\alpha = 45°$ for undrained failure

$$F - \frac{E_d}{E_i} = \frac{Cu}{\gamma h_1}\left(2 + \frac{R}{2h_2 St}\right) \qquad (4.10)$$

where St is the soil sensitivity. Thus for a case where earthflow would develop ($F = 1$), the retrogressive distance can be estimated as

$$R = 2h_2 St\left(\frac{\gamma h_1}{Cu} - 2\right) \qquad (4.11)$$

The application of this type of analysis to earthflows in the Champlain Sea sediments of Canada is discussed in some detail by Carson (1977). In many cases where earthflows have occurred the values of sensitivity measured by the field vane test have exceeded 100, but it is likely that the shearing resistance in the remolded shear layer is actually higher than the remolded vane value (Ladanyi et al. 1968). For Champlain Sea Clay a value of Cu/Cu_R of about 8 would appear appropriate and in many cases it is found that $h_2 \simeq H/2$ giving

$$\frac{R}{H} = 4\left(\frac{\gamma H}{Cu} - 4\right) \qquad (4.12)$$

From case studies of earthflows Mitchell and Markell (1974) propose an earthflow chart in which, using $N_s = \gamma H/Cu$,

$$3(N_s - 4) \leq R/H \leq 20(N_s - 4) \qquad (4.13)$$

Although Equations 4.11 and 4.13 can be used to give a preliminary estimate of the potential retrogressive distance R for slope failures in earthflow-prone areas, it is recommended that air photo studies be used to correlate results in

119

a particular area. Mitchell (1978), for example, suggests that the following relations are applicable to the Ottawa–St Lawrence lowland areas:

$$R \le 100 \text{ m} \qquad \text{for } N_s \le 5$$

$$R \le 100(N_s - 4) \text{ m} \qquad \text{for } N_s > 5$$

Rapid earthflows have developed in bonded sediments during adjacent pile-driving operations (Broms 1978, Carson 1977). Strain softening and high pore-water pressures caused by the vibrations and lateral displacements due to pile-driving are considered to have triggered these earthflows. Rapid earthflows have also developed in saturated sands and silts due to liquefaction under earthquake or blast accelerations (see, for example, Seed & Wilson 1967, Seed 1968). Large rapid submarine earthflows, often triggered by earthquakes or wave action, are also known to occur (Terzaghi 1956, Bea & Audibert 1980).

Figure 4.5 depicts a slow earthflow or mudflow where the mechanism of flow involves sliding on a number of subparallel planes with movements generally developing over a long period of time under high seasonal groundwater conditions due to heavy rainfall or snow melt. In such cases c' will approach zero on the sliding planes and the safety factor can be estimated as

$$F = \frac{\tan \phi_3' (1 - r_u \sec^2 \alpha_3)}{\tan \alpha_3 + F\alpha_3/W_3 \cos \alpha_3} \qquad (4.14)$$

where

$$F\alpha_3 = W_2 \sin \alpha_2 + F\alpha_2 - \left[\tan \phi_2' \left(W_2 \cos \alpha_2 - \frac{u_2 L_2}{\cos \alpha_2} \right) \right] \cos | \alpha_2 - \alpha_3 |$$

$$F\alpha_2 = W_1 \sin \alpha_1 - \left[\tan \phi_1' \left(W_1 \cos \alpha_1 - \frac{u_1 L_1}{\cos \alpha_1} \right) \right] \cos | \alpha_1 - \alpha_2 |$$

Careful site investigation and testing is required to determine the location of sliding planes and to measure the value of ϕ' on these planes. In many cases the shearing resistance on these planes is close to the residual values for the materials involved and slope drainage (reducing u) or soil reinforcement may be useful in reducing soil movements (VanDine 1980). Slickensided failure surfaces as shown on Figure 4.9 are often evident when a residual strength has been attained. In cases where ϕ' is well above the residual value on any plane, further movement may reduce the shearing resistance and there is a possibility of accelerating movement developing during exceptionally heavy rainfall.

Figure 4.9 Slickensided failure surface near toe of Drynoch landslide. (Photo courtesy of British Columbia Ministry of Transportation and Highways.)

4.2.5 *Influence of stratigraphy on slope stability*

Weak layers or pervious zones in slopes can have a marked effect on the stability of a slope. In some cases the weaknesses are not readily apparent from site investigation information. Shales are particularly prone to progressive deterioration after exposure – due to bentonite seams, shale mylonites and associated slickensided fissures (see, for example, Scott & Brooker 1968, Deere & Patton 1971). Computer solutions which analyze sliding on non-circular surfaces are generally recommended when weak or pervious layers are present in a slope profile. Most computer solutions do not, however, consider seepage forces as active driving forces and in some cases a simplified analysis which includes these forces is necessary. In other cases it is convenient to use active and passive earth pressure concepts in obtaining closed-form approximate solutions to slope stability problems. Two cases where these concepts are useful in estimating slope stability are considered in this section.

Figure 4.10a shows a situation where a dry granular material is rapidly stockpiled on a subgrade containing a soft cohesive layer which outcrops a short distance from the toe of the pile. In this case the safety factor against lateral sliding on the weak layer can be approximately evaluated as

$$F = \frac{Cu(L + H \cot \beta)}{P_A} = \frac{Cu(L + H \cot \beta)}{\frac{1}{2}\gamma H^2 \tan^2(45° - \phi/2)} \tag{4.15}$$

Figure 4.10b shows a situation where saturated intercalated sand or silt lenses form a semicontinuous zone of more pervious material within a less pervious sediment containing clay layers on which lateral sliding can develop. Under heavy rainfall, snow melt or ponded surface waters the pervious zone can support hydrostatic water pressures. The safety factor in this case can be approximated as

$$F = \frac{Cu\,R}{\tfrac{1}{2}\gamma' H^2 \tan^2(45° - \phi'/2) + \tfrac{1}{2}H^2\gamma_w} \tag{4.16}$$

Although the condition idealized on Figure 4.10b is not likely to occur in nature, a condition approaching this idealization may develop during a retrogressive landslide where stress release promotes lateral movement and vertical cracking in stiff sediments containing clay layers and intercalated sands. It is interesting to note that Equation 4.16 indicates that retrogression to distances of twice the slope height $(R = 2H)$ can develop even when $\gamma H/Cu$ is less than 5. Photos of large earthflow-like craters which were formed in sandy soils by the processes of removal of materials due to internal water pressures (sapping) have been published by Mollard (1973), and Mitchell and Klugman (1979) discuss the formation of internal erosion craters in banded sediments. Blondeau and Queyroi (1976) report a case of failure of a temporary slope in a plastic clay due to rapid pore-water pressure equilibrium as a result of an underlying drainage layer.

The above examples are intended to demonstrate the necessity of detailed investigations of stratigraphy in slope stability analysis but also show that elementary soil mechanics considerations should not be neglected when computer solutions are available. Finding the weakest link in a given situation is still the job of the engineer. Some field testing techniques for investigating stratigraphy and *in-situ* properties are discussed by Sowers and Royster (1978).

4.2.6   *Influence of material behavior on slope stability analysis*

The classic alternative to an upper-bound or limiting equilibrium analysis is a lower-bound analysis, where only one point in a statically admissible stress distribution is allowed to attain yield or failure. Application of this type of analysis to the stability of earth masses is restrictively conservative in most cases. As stress analysis techniques improve, however, the results of stress–strain analysis of slopes can be very useful in evaluating the mobilized strength of the soil mass at failure (see, for example, Lo & Lee 1973). Although time effects can also be included in the analysis, the basic approach is outlined on Figure 4.11. Stress analysis is used to estimate the location and extent of overstressing in the soil mass (yielded zone), and the overstressed soil is assigned a reduced post-peak strength in the stability analysis. Law

(a) Graben formation

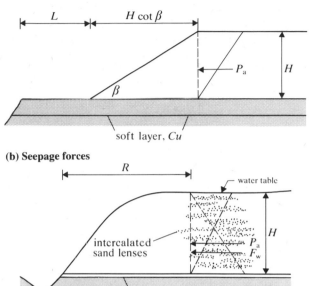

(b) Seepage forces

Figure 4.10   Active and seepage forces.

and Lumb (1978) offer an alternative approach whereby local failure and strength reduction on a trial arc is evaluated by considering the statics of the development of interslice forces.

Progressive failure is a term generally used to describe the processes of softening and reduction in shearing resistance in slopes after primary equilibrium has been achieved. Two distinctly different types of progressive failure are recognized in geotechnical literature. Skempton (1964) describes rotational slope failures which developed in heavily overconsolidated clays up to 150 years after the slopes were excavated. The general softening of a mass of soil in these slopes appears to have developed due to passive failure (Singh *et al.* 1973), and the long-term strength of the mass approaches a fully softened strength which can be obtained by testing samples that are remolded and then reconsolidated and tested as normally consolidated samples (Skempton 1970). The lowest normally consolidated or fully softened strength can be approximated from Equations 2.2 and 2.9 as

$$c' = 0$$

and

$$\phi' = \sin^{-1}[(0.11 + 0.37PI)/(0.89 - 0.37PI)]$$

123

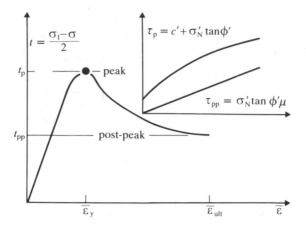

stress analysis used to obtain yield zones

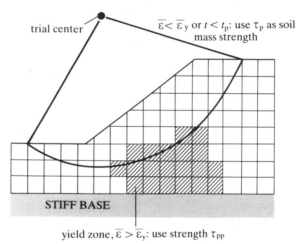

yield zone, $\overline{\varepsilon} > \overline{\varepsilon}_y$: use strength τ_{pp}

Figure 4.11 Modification for yield.

It is recommended that permanent slope cuttings in heavily overconsolidated clays ($OCR \geq 10$) of low or normal activity be designed on the fully softened strength. Further experience with progressive failures of this type has been documented by Palladino and Peck (1972).

Bjerrum (1967b) describes and analyzes translatory sliding in natural slopes in heavily overconsolidated residual soils. In these cases stress release due to slope formation caused creep movements to occur on pre-existing weak planes, or caused the development of slickensided surfaces in layers of active weathering and swelling. The mobilized strength in these planes was calculated to be the residual strength of the material. Further details of residual soil failures are provided by Deere and Patton (1971). It is recommended that residual strength be used in analyzing the effects of slope cuttings in active residual soils (generally soils containing swelling minerals) and in cases where weak slickensided planes exist due to previous movements (Kenny 1967a, Scott & Brooker 1968).

Partially saturated materials are often stable by virtue of the capillary suctions existing within the mass. Any construction activity that could result in increasing the degree of saturation of the mass would also result in decreasing the mass strength. Partially saturated slopes should be designed on the basis of fully saturated strength data.

In closely jointed or badly fractured earth materials the rotational sliding mechanism is normally assumed, although a computer simulation model to describe the movements of discrete blocks has been developed by Cundall (1974). When c', ϕ' strength parameters are used to describe these materials, the strength testing should be designed to account for the effects of sample size (Lo 1970, Marsland 1971) and failure by mass dilation (Ladanyi & Archambault 1970, Eden & Mitchell 1970).

The development of landsliding in frozen and thawing soils is discussed by McRoberts and Morgenstern (1974). Testing considerations and typical strength parameters for various types of materials are contained in Chapter 2 and in Wu and Sangrey (1978). Undisturbed test samples should be obtained by sampling in slopes close to the theoretical critical failure planes (or arc) and the structure and fabric of samples should be closely examined.

4.3 Design charts for slopes in homogeneous materials

Homogeneous soils are defined for analytical purposes as soils in which there are no continuous weak planes on which sliding can develop and in which the groundwater pressures can be reasonably estimated, as required, by an average r_u value. In such cases rotational sliding is the most probable failure mechanism and design charts can be produced from computer solutions. Most available design charts have been produced using the Bishop simplified analysis.

4.3.1 Evaluation of critical centers and r_u for design charts

The average r_u for a homogeneous groundwater condition can be evaluated from a flow net or from the position of the top flow line (phreatic surface) with respect to the location of the critical failure arc. For homogeneous soil conditions the location of the critical center may be estimated with reference to Figure 4.12. For low values of r_u (<0.3) the critical center is well approximated by

$$H_c = H \cot \beta \,(0.6 + 2 \tan \phi')$$

$$(4.17)$$

$$L_c = H \cot \beta \,(0.6 - \tan \phi')$$

When the r_u value is expected to exceed about 0.3, the location of the critical center is less dependent on tan ϕ' and is influenced to a greater degree by the value of c' such that deeper critical circles prevail. The information on Figure 4.12a can be used to estimate the location of the critical center in such cases. Provided that tan ϕ' is greater than 0.1, however, the critical arc will

(a) Approximate locations of critical centers for F = unity

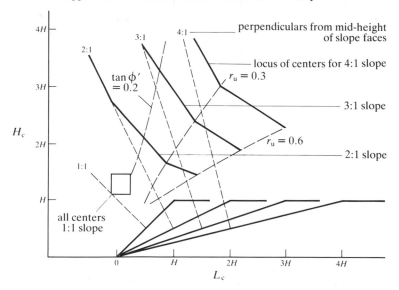

(b) Evaluation of r_u for use with design charts

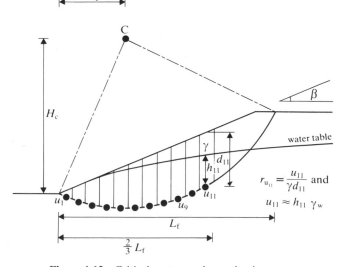

$$r_{u_{11}} = \frac{u_{11}}{\gamma d_{11}} \text{ and}$$

$$u_{11} \approx h_{11} \gamma_w$$

Figure 4.12   Critical centers and r_u evaluations.

126

pass through the toe of the slope. Once the approximate location of the critical arc is found, it is recommended that individual values of $r_u = u/\gamma d$ be obtained for about 10 equal-width slices composing the lower two-thirds of the potential failure mass. This is done using a flow net or, for relatively flat slopes (30° or less) without artesian or drawdown pressures, using the assumption that $u = h\gamma_w$, where h is the vertical distance from the base of the slice to the phreatic surface (as shown on Figure 4.12b). This latter assumption results in a marginal overestimate of the water pressure u in a homogeneous material. The average r_u value for the slope is then calculated as

$$r_u(\text{avg}) = \sum_1^N (u/\gamma d)/N \qquad (4.18)$$

where N is the number of equal-width slices composing the lower two-thirds of the slope.

In evaluating r_u it must be remembered that rainy seasonal groundwater pressures can be considerably higher than those measured during the rest of the year. Also slope surface infiltration, during heavy or extended rainfall, can increase the groundwater pressure within a few hours by saturating the material above the water table. It is well known that most landslides occur during wet seasons and often during heavy rainfall. These factors must be considered in evaluating r_u. Some recommendations are contained on Figure 4.13 for three common flow conditions. Figure 4.13c also shows the suggested locations for the minimum number of piezometers (P_1 to P_4) necessary to establish the existence of artesian or drawdown conditions.

Figure 4.12 and Equations 4.17 provide an indication of how the location of the critical center varies with slope geometry, soil properties and groundwater pressure. Typical safety factor contours of trial center locations are shown on Figure 4.14 for given slope conditions. Such contours are often used to indicate the risk of upland facilities being involved in a landslide (providing no retrogressive sliding is anticipated).

4.3.2 Design charts for rotational sliding

Provided that the slope conditions can be accurately represented by a constant slope angle (β), homogeneous strength parameters and a relatively homogeneous average r_u value, chart solutions are as accurate as computer solutions. Design charts are also useful for rapid preliminary analyses to determine whether existing slopes are sufficiently critical to warrant a more detailed analysis.

The earliest published design chart (Taylor 1937) was for materials having a constant strength. Later charts incorporated strength increase with depth (Gibson & Morgenstern 1962) and the more general c', ϕ' strength conditions (Bishop & Morgenstern 1960).

(a) Parallel flow, no slope seepage

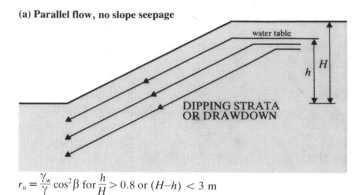

$$r_u = \frac{\gamma_w}{\gamma} \cos^2\beta \text{ for } \frac{h}{H} > 0.8 \text{ or } (H-h) < 3 \text{ m}$$

(b) Horizontal flow, full slope seepage

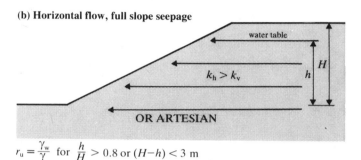

$$r_u = \frac{\gamma_w}{\gamma} \text{ for } \frac{h}{H} > 0.8 \text{ or } (H-h) < 3 \text{ m}$$

(c) Parabolic top flow line

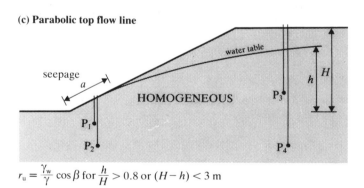

$$r_u = \frac{\gamma_w}{\gamma} \cos\beta \text{ for } \frac{h}{H} > 0.8 \text{ or } (H-h) < 3 \text{ m}$$

Figure 4.13 Design values of r_u.

The Bishop and Morgenstern (1960) charts and extensions to these charts (O'Connor & Mitchell 1977) are in general use for permanent slope stability analysis and design. These charts provide numerically accurate values of safety factor ($+0.01$) for the Bishop simplified equation in the form

$$F = m - n(r_u)$$

128

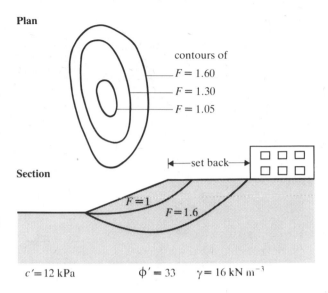

Plan

contours of

$F = 1.60$

$F = 1.30$

$F = 1.05$

←—set back—→

Section

$F=1$

$F=1.6$

$c' = 12 \, \text{kPa}$ $\phi' = 33$ $\gamma = 16 \, \text{kN m}^{-3}$

Figure 4.14   Safety factor contours.

by providing values of m and n for selected values of $c'/\gamma H$, β, ϕ' and D (as well as providing an equivalent r_u which is used to estimate the critical depth of failure). Use of these charts generally requires iteration between charts of constant $c'/\gamma H$, however, and this increases the risk of error during the processes of computation. The general availability of digital computers also reduces the need for such accurate charts since critical situations can be analyzed in much greater detail on the computer, using a more detailed and accurate description of the relevant parameters.

Hoek and Bray (1974) introduced a new format for plotting permanent design charts, as exemplified on Figures 4.15 through 17. The charts published by these authors were intended for rock slope design, where steep slope angles are normal and where the groundwater flow regime is often not classified in the same detail as used for soil slopes. As a result the original published charts are not accurate in the range applicable to soil slopes. Figures 4.15 through 17 have been produced from computer solutions of Equation 4.5 and an allowance for tension cracks at the crest of the slope has been made for slopes of $\cot \beta < 1$. Since both $\cot \beta$ and F vary linearly with r_u, linear interpolation of either F or the slope ($\cot \beta$) between the three charts can provide accurate design for intermediate groundwater flow conditions.

The advantages of the chart format on Figures 4.15 through 17 are operational simplicity, fewer charts with less iteration and a check on the value of F (two values, which should be within 0.05 are obtained in each case). Two examples will serve to illustrate the use of these charts.

Example 1   Given $c' = 10 \, \text{kPa}$, $\phi' = 36°$, $r_u = 0.25$, $H = 16.8 \, \text{m}$, $\gamma = 18 \, \text{kN m}^{-3}$,

129

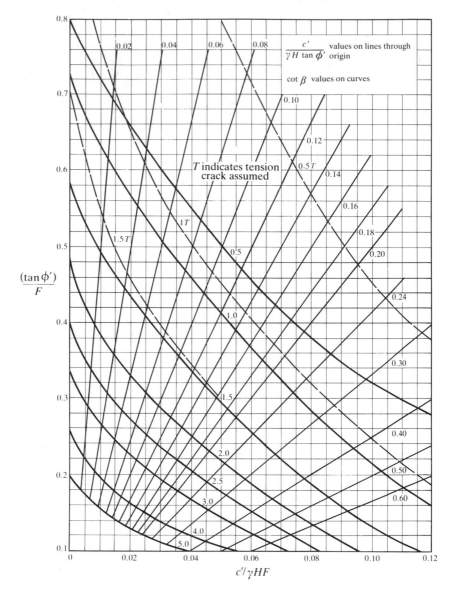

Figure 4.15 Design chart for rotational sliding. $r_u = 0$.

cot $\beta = 3$; find F. Calculate $c'/\gamma H \tan \phi' = 0.045$, $(\tan \phi' = 0.727)$ and enter charts on this line to find intersection with cot $\beta = 3$. For $r_u = 0$ (Fig. 4.15) obtain $\tan \phi'/F = 0.258$ and $c'/\gamma HF = 0.012$, giving $2.73 \le F \le 2.28$ (avg $= 2.78$). For $r_u = 0.3$ (Fig. 4.16) obtain $\tan \phi'/F = 0.359$ and $c'/\gamma HF = 0.016$, giving $2.03 \le F \le 2.04$ (avg $= 2.03$). These values are then iterated to give $F = 2.78 - (0.25/0.3)(2.78 - 2.03) = 2.16$.

130

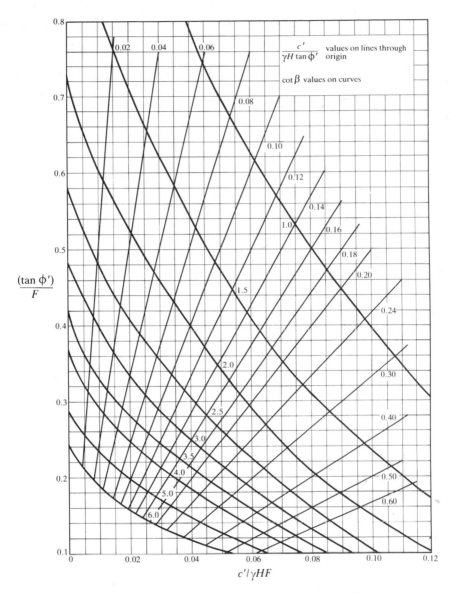

$\dfrac{(\tan \phi')}{F}$

$c'/\gamma HF$

Legend inside chart:
$\dfrac{c'}{\gamma H \tan \phi'}$ values on lines through origin

$\cot \beta$ values on curves

Figure 4.16 Design chart for rotational sliding, $r_u = 0.3$.

Example 2 Given $c' = 15$ kPa, $\phi' = 28°$, $r_u = 0.4$, $H = 10$ m, $F = 1.25$, $\gamma =$ 17 kN m⁻³; find β. Calculate $c'/\gamma HF = 0.071$, $\tan \phi'/F = 0.426$, $c'/\gamma H \tan \phi'$ $= 0.167$. Figure 4.17 gives $\cot \beta = 2.25$ ($r_u = 0.6$) and Figure 4.16 gives $\cot \beta$ $= 1.30$ ($r_u = 0.3$). Iterating these values, $\cot \beta = 1.30 + (0.1/0.3)(2.25 - 1.30) = 1.62$. Use $\cot \beta = 1.6$ ($\beta = 32°$).

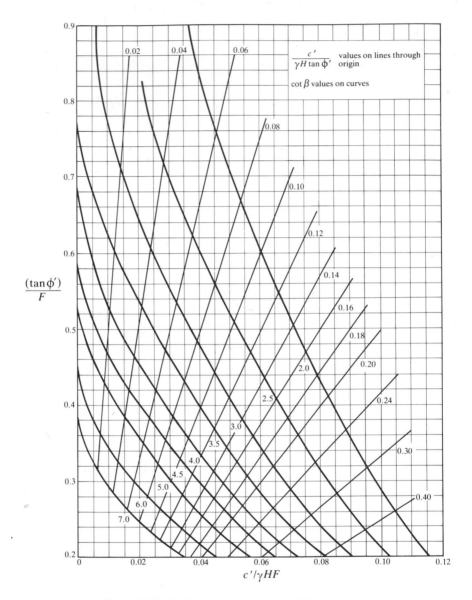

Figure 4.17 Design chart for rotational sliding, $r_u = 0.6$.

The charts on Figures 4.15 through 17 are sufficient to design permanent slope cuttings or to evaluate the safety factor of existing slopes in homogeneous earth materials. When safety factors of homogeneous natural slopes are found to be less than 1.3, a more detailed computer analysis based on measured groundwater pressures is warranted. Indeed, whenever large sums of money are expended on site investigation and laboratory testing, it seems

132

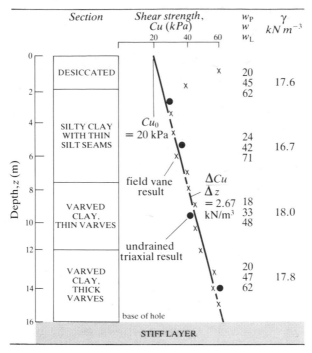

Figure 4.18 Typical borehole log.

reasonable to use a computer solution that is capable of using the detailed data rather than a chart solution that requires averaging of material properties.

4.3.3 Stability numbers for temporary slopes in fine-grained soils

If an excavation is opened for a short time period in a fine-grained saturated soil, the short-term undrained strength of the soil can be used in design. The undrained strength profile is normally obtained using field vane tests supported by undrained triaxial testing of samples from selected depths in the profile. Figure 4.18 shows a typical result. Because of the high probability of tension cracks developing in the stiffer desiccated crust, it is usual to extend the best-fit line through the test results to characterize the strength profile as

$$Cu = Cu_{[o]} + \left(\frac{\Delta Cu}{\Delta z}\right) z$$

as shown on Figure 4.18. If the strength increases with depth sufficiently such that

$$\frac{\Delta Cu}{\Delta z \gamma} \geq 0.08$$

133

the critical arc will pass through the toe of a slope cutting and the short-term safety factor is given by

$$F = N_s \frac{Cu_{[0]}}{\gamma H} + N_0 \frac{\Delta Cu}{\Delta z \gamma} \tag{4.19}$$

where H is the height of the slope (depth of excavation) and N_s and N_0 are given, for practical slope angles, on Table 4.1.

When the strength increase with depth is marginal,

$$\Delta Cu/\Delta zy < 0.08$$

the soil is considered to be a constant-strength material (use the average strength) and the critical circle will extend down to a stiff base. For example, an excavation of 10 m depth (slope height, $H = 10$ m) in a constant-strength silty clay material overlying a stiff till at 16 m depth would be characterized by a depth factor, $D = 16/10 = 1.6$. Excavations to rock are often carried out for construction or mining purposes and these have $D = 1$. In all these constant-strength cases the factor of safety, F, is obtained by equating $Cu/\gamma HF$ to the relevant constant listed on Table 4.1. Figure 4.19 shows some cases where the relevant value of D is easily determined. It is unusual for a soil to have an undrained strength that is constant with depth (due to a general decrease of void ratio with depth), however, and Equation 4.19 would normally be used for temporary slope design. A more detailed alternative to Equation 4.19 is available using plotted charts (Hunter & Schuster 1968) but the stability factors N_s and N_0 on Table 4.1 are considered sufficiently accurate for temporary slope design. A relatively high safety factor should be employed in the design of temporary slopes because of the uncertainties involved in both the undrained strength measurements and the time required for pore-water pressure equilibrium. Safety factors generally in excess of 1.5 are used but further detailed recommendations are contained in Section 4.5.

Table 4.1 Stability numbers for temporary slopes in homogeneous soils.

Slope angle, β (deg)	Increasing strength		Constant strength			
				$Cu/\gamma HF$ for depths to		
	N_s	N_0	$D = 1$	$D = 1$–1.5	$D = 1.5$–2	$D > 2$
18.5	6.7	7.7	0.11	0.14	0.16	0.18
25	6.3	6.4	0.13	0.15	0.17	0.18
30	5.9	5.5	0.14	0.16	0.18	0.18
40	5.3	4.6	0.16	0.17	0.18	0.18
50	4.8	3.8	0.18	0.18	0.18	0.18
60	4.3	3.3	0.19	0.19	0.19	0.19
70	3.8	2.8	0.21	0.21	0.21	0.21

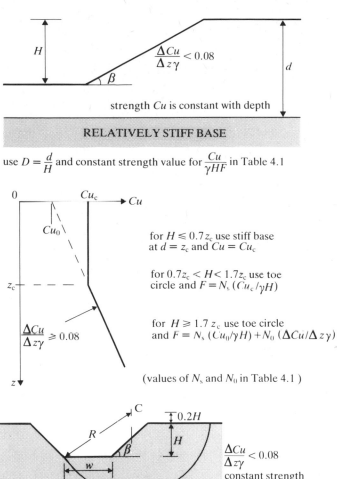

$$D \approx \frac{[1.4H^2 + (w + H \cot \beta /2)^2]^{\frac{1}{2}}}{H} - 0.2$$

Figure 4.19 Typical short-term situations.

4.4 Crest loadings, dynamic loadings, submergence and drawdown

The preceding analysis and design charts consider only static gravitational loadings on slopes in air. In some cases earth slopes will be subjected to crest loadings, dynamic loadings, submergence or rapid drawdown (removal of water from a canal or reservoir in a relatively short period of time). In such cases the slope must be designed to withstand these operational conditions.

135

4.4.1 Crest loadings on slopes

Perhaps the only significant legitimate crest loading on a critical high soil slope ($H > 10$ m) is a dragline as used in cut-and-cover mining operations. Large buildings, above-ground swimming pools, embankments, waste dumps and the like should not be placed near the crest of a slope unless the slope is composed of exceptionally strong and well drained materials (rock or granular soil), or can be suitably stabilized to accept this crest loading. Setbacks (the allowable proximity of a structure to the crest of a slope) are often established with reference to the locations of trial circles giving a particular safety factor as shown on Figure 4.14. This method is only recommended for homogeneous insensitive materials where the slope is stable with respect to geometry and strength considerations. It may be noted that houses with basements do not normally create any net loading (provided that the excavated soil is removed from the site) but hazardous terrain and actively eroding slopes should be avoided in residential site selection. Using a computer program any crest loading can be incorporated, with appropriate γ and strength (if desired), in the slope geometry. Using design charts the following two approaches can be used:

(a) Increase the slope height to include the crest loading – a 2 m deep pool is, for example, equivalent to $(2\gamma_w/\gamma)$ m of soil;
(b) Evaluate the driving moment due to the crest loading within the limits of the critical arc as a fraction, f, of the slope material driving moments and divide the chart value of F by $(1 + f)$.

The dragline problem is often a short-term problem since the cutting is mined at such a rate that the slope is not exposed for too long a period of time. For the simple situation where the overburden can be represented as a constant-strength material, the chart on Figure 4.20 shows the effect of a large dragline on slope stability. Usually the overburden strengths in such a situation are not so easily described and a computer solution is warranted. Figure 4.20 does indicate, however, the significant effect that the mobilization of such heavy equipment along a slope crest can have on the safety factor. The practice of dumping snow from city streets on the crest of river banks or ravine slopes can initiate slope movements, not only because of the weight of the snow but also because of melt infiltration and possible chemical effects due to salts.

4.4.2 Earthquake and other dynamic loadings

For materials not susceptible to liquefaction or other structural breakdown due to vibrations the effect of earthquake loadings can be considered in terms of equivalent static forces, as shown on Figure 4.21, where both the

(a)

$$D_w = \frac{\text{dragline total weight}}{\text{tub or track bearing width}} \quad \left(\frac{kN}{m}\right)$$

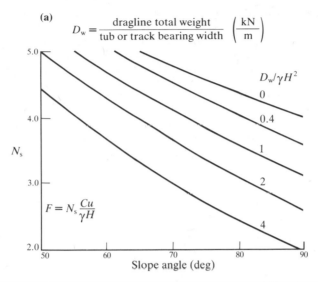

$D_w/\gamma H^2$

0

0.4

1

2

4

N_s

$$F = N_s \frac{Cu}{\gamma H}$$

Slope angle (deg)

Figure 4.20 Dragline crest loading on a cohesive slope. (a) Chart showing effect of a large dragline on slope stability. (b) Dragline operating on a mine slope. (Photo by D. Stone.)

gravitational and earthquake accelerations are applied to the mass to calculate the maximum driving forces $(g + a_v)M$ and minimum resisting forces (since σ_N' is reduced by the acceleration a_h). Earthquake loading is cyclic (approximately 1 to 3 Hz) and the safety factor will theoretically increase and decrease causing sympathetic downslope ground movements. For short-duration quakes this analysis is considered conservative because the material

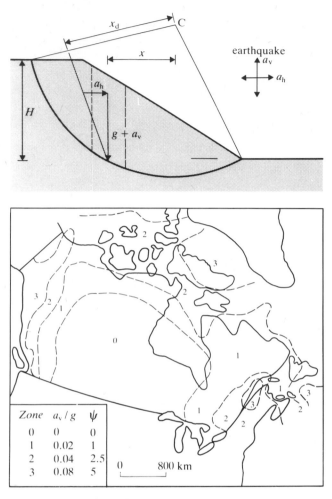

Figure 4.21 Earthquake design.

will have little time to respond to the loading. For longer-duration earthquakes (several minutes) the analysis is more realistic since progressive movements may develop a slip mechanism. Nevertheless low safety factors can normally be employed with this static analysis. Earthquake accelerations may be programmed in computer solutions and it is normal procedure to assume that the pore-water pressures remain constant during earthquake loading on insensitive materials. For materials such as silts and sands (and, possibly, very sensitive clayey silts) with a high degree of saturation, a partial or complete loss of strength (liquefaction) can occur after several cycles of stress reversals. This particular problem has been the subject of considerable research, and design procedures, including dynamic computer analysis, have been developed (Seed 1968, Dunlop *et al.* 1968).

138

To obtain approximate safety factors for homogeneous soils under earthquake accelerations an equivalent static slope having

$$H_e = H(1 + a_v/g)$$

and

$$\beta_e = \beta + \psi$$

can be analyzed using design charts. Figure 4.21 provides recommended factors for this type of approximate analysis in the three earthquake zones of Canada. More detailed recommendations with regard to probable earthquake accelerations in various areas are available from national building codes and research publications (see, for example, Whitham *et al.* 1970, Milne & Rogers 1972, Basham *et al.* 1979).

Blast vibrations are normally of high frequency and low duration and have little effect on the stability of slopes provided that good blasting control is exercised (Hoek & Bray 1974) and that the material is not susceptible to liquefaction. Susceptible fill materials should be well drained or compacted dry of optimum to reduce or eliminate the risk of liquefaction failures. Slope movements can develop in sensitive soils or saturated silty soils under low-frequency long-duration dynamic loadings such as those created by pile-driving. This problem is discussed with reference to excavation slopes by Crooks *et al.* (1980).

4.4.3 *Submergence, drawdown and wave action*

When free water overlies a slope face it is called a partially submerged or, if completely under water, a submerged slope. If the submerging water is quickly drained away, the slope is said to have been subjected to rapid drawdown – this can be full or partial drawdown. Each year, during flood flows, river valley slopes are subjected to submergence and drawdown but, in most cases, this effect is small compared to the effect of bank erosion in changing the stability of the slope. When reservoirs or canals are created, however, the submergence and drawdown effects can control the design of impoundment slopes.

The driving forces promoting failure of a full submerged slope are calculated from the submerged unit weight of the material (hence, these are reduced by approximately 50%, for a typical soil, from the non-submerged condition). The resisting forces will be the same in both cases for a purely cohesive material but may change, for a frictional material, depending on the groundwater conditions assumed for comparison. For example, a hydrostatically saturated slope in air has $r_u = \gamma_w/\gamma$ and the same condition exists when fully submerged; a drained slope ($r_u = 0$), however, would have almost

139

twice as much resistance to sliding in air than when submerged since γ_w/γ is approximately equal to 0.5 for a typical soil. The following general rules of thumb can be stated from these considerations:

(a) The stability of a given slope will generally be greater if the slope is submerged than if it exists in air.
(b) The stability of a given slope will decrease during drawdown.

One exception to the first statement is when, during early reservoir filling, the natural groundwater flow may be backed up in the reservoir slope creating higher groundwater pressures in the slope. Flow nets should be constructed for different levels of reservoir filling using regional ground-water flow data and computer solutions to examine this possibility. The only other exception is when submergence causes ground water to wet previously dry materials that expand or lose strength when wetted (partly saturated silts, expansive clays or fine-grained material infilling discontinuities in rocks). Laboratory and field testing should be carried out to examine these possibilities.

When a slope is fully submerged its safety factor may be obtained from charts or formulae by using $c'/\gamma'H$ in place of $c'/\gamma H$ or $W' = W(1 - \gamma_w/\gamma)$ in place of W and equating r_u (or u) to zero. These substitutions calculate both driving and resisting forces on the basis of submerged unit weight and no groundwater flow, the correct conditions for a fully submerged slope. With the exceptions due to geological factors (as noted above), interpolation between the original conditions and the fully submerged condition can be used to estimate the safety factor of a partially submerged slope.

When a fully submerged slope is subjected to rapid drawdown, the water contained in the slope tends to flow to the slope face, a condition of horizontal flow given by $r_u = \gamma_w/\gamma$. Thus the extreme condition of full rapid drawdown can be analyzed using design charts with $r_u = \gamma_w/\gamma$. An analysis which includes the effects of soil properties on the pore-water pressure parameter has been presented, with design charts, by Morgenstern (1963) for conditions of partial drawdown to various reservoir levels. Drawdown and submergence charts can, however, be prepared by the methods outlined on Figure 4.22. A chart similar to Figure 4.22 is prepared by the following steps:

(1) The safety factor for the fully submerged slope ($L = 0$) is calculated and plotted (point S).
(2) The factor of safety for full rapid drawdown ($L = H$) is calculated and plotted (point R).
(3) Project a horizontal line from R to $L = 0.5H$ and join this point (point a) to point S.
(4) Construct line bc between the quarter point intersections of lines aR and aS.

140

(5) Construct the drawdown curve from S to R tangent to bc as shown.
(6) Construct the submergence lines as shown on Figure 4.22. This construction assumes little to no increase in safety factor during the first 25% of submergence due to groundwater pressure increases.

Submerged and partly submerged slopes are subject to the erosive action of waves, currents and ice forces. Canal slopes are normally lined with impervious membranes (if ground water is relatively low) or protected by rip-rap or dolos blocks on filter media (when ground water is relatively high). Erosion protection is needed above and below the waterline when a steep submerged slope is composed of easily erodible materials. When the

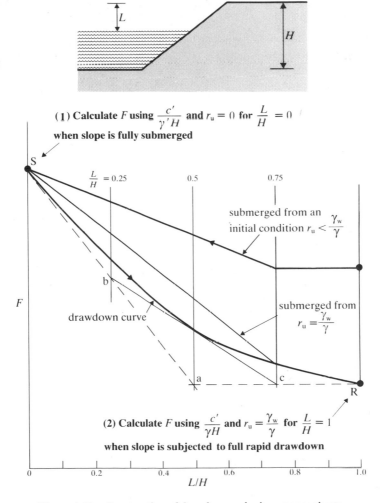

Figure 4.22   Construction of drawdown and submergence charts.

141

slope angle is fairly low, slope erosion can be allowed to develop a stable beach in suitable materials.

In many cases of offshore (submarine) prodelta sediments it has been shown that wave action from storm waves can cause large underwater landslides in underconsolidated and normally consolidated sediments (Bea & Audibert 1980, Algamor & Wiseman 1977). The mechanics of wave-induced submarine landsliding has been investigated from a static upper-bound analytical approach by Henkel (1970b), from a finite element approach by Wright and Dunham (1972) and from dynamic similitude modeling techniques by Mitchell et al. (1972b) and Mitchell and Hull (1974).

4.5 Recommended design factors of safety and procedures

The growing volumes of literature on statistical analysis in geotechnical engineering, some of which contain the suggestion that probability of failure is more meaningful than factor of safety (Wu & Kraft 1970, Yong et al. 1977), reinforce the general feeling that the design (or safe) value which is selected for F should be chosen with due consideration for the confidence established for both the analysis and the strength test results. Where this confidence is established by field performance, low safety factors can be justified. When one considers that the volume of earthwork associated with earth cuttings or embankments is proportional to the cotangent of the slope angle, there is a substantial economic incentive to strive for low safety factors in large earthworks.

A temporary slope is stable by virtue of the pore-water pressure reduction induced by a reduction of the total confining stresses as material is excavated. A saturated fine-grained soil can maintain these pressure reductions for some time due to its low permeability, but it is acknowledged that a slope will eventually fail if constructed on the basis of the temporary slope design. For many soils a temporary slope safety factor of about 2 will yield a long-term factor of safety of about 1 (this can be checked, for a particular soil, if the applicable values of c' and ϕ' are known from drained test results). On this basis, and considering that the long-term condition will be achieved in about three months for silty clays (or varved clays) and in about six months for intact clay soils, the following safety factors are recommended:

$$F = 1.3 + 0.2M \qquad \text{for silty clays}$$
$$\text{(4.20)}$$
$$F = 1.3 + 0.1M \qquad \text{for intact clays}$$

where M is the number of months that the excavation will be open.

These recommendations may be modified on the basis of field experience and risk factors to provide either greater safety (higher F) or greater excavation economy (lower F).

142

Factor of safety recommendations for permanent slope design are listed on Table 4.2. These recommended safety factors should be increased by 0.25 where there is limited material testing and by an additional 0.25 where there is limited construction control (in the case of dikes and dams). Thus, a small impoundment dam in a low-risk situation could be constructed using $F = 1.3$ with good engineering and supervision while $F = 1.8$ is recommended if there is a lack of engineering design and supervision. For a dam 10 m in height, the construction cost difference would be in the order of $500 per meter of dam length and the cost of good engineering design and construction supervision could be recovered in the first 50 to 100 m of dam length.

The steepest natural slopes in actively eroding landscapes exist at a safety factor close to unity and some of these may be stable for decades while failure is triggered in neighboring slopes by climatic, erosional or stratigraphic factors. Erosion and landslide patterns can be observed, as discussed in Chapter 1, using air photos. It is commonly found that slope creep movements develop in many earth slopes where the calculated safety factor is less than about 1.25 (Mitchell & Eden 1972). Safety factors in excess of this value should be used for permanent earth cuttings in materials where creep movements are evident in natural slopes.

The factor of safety recommendations on Table 4.2 are based on the assumption that site investigation and testing procedures have identified the most probable mechanism of failure and the applicable strength envelope. When failure can develop in weak clay layers, samples of this clay should be subjected to multiple shear reversals in a shear box to investigate the

Table 4.2 Typical safety factors.

Slope description and design conditions	High risk (loss of life or severe damage from failure)	Low risk (no loss of life and moderate damage from failure)
permanent slope in geologically stable material, all conditions	1.3	1.15
permanent slope in geologically metastable materials, all conditions	1.5	1.25
non-impoundment dikes and embankments, all conditions	1.3	1.3
impoundment dams		
(a) end of construction	1.3	1.3
(b) normal operation	1.5	1.3
(c) rapid drawdown	1.3	1.1
(d) earthquake loadings	1.2	1.1
(e) earthquake in combination with (a), (b) or (c)	1.1	1.0

A geologically metastable material is intended to refer to a material susceptible to earthflow or where low safety factors may lead to creep movements and progressive softening.

143

Table 4.3 Summary of slope stability design and construction.

Item	Summary notes
site investigation	Map and air photo stability survey. Field testing to investigate stratigraphy using vane, penetration and joint mapping. Undisturbed sampling to obtain materials from weak layers or close to potential sliding surfaces (*in-situ* tests as applicable). Piezometer installations to obtain groundwater pressure distribution
material testing (material identification and classification required for correlations)	Drained triaxial testing of undisturbed samples in the correct normal stress range for c', ϕ'. Large-diameter tests or large shear box tests on blocky or fractured materials. Direct shear testing of undisturbed weak planes or joints in direction of potential movement in correct normal stress range. Softening, swelling and residual strength testing in heavily overconsolidated and residual materials
analysis (back-analysis of any previous slides to confirm mechanisms and strength)	Determine potential failure mechanisms from site investigation and testing details as well as previous experience (literature). Analyze potential failure mechanisms using charts, equations or computer analysis as required to obtain safety factor or design slope angles. Long-term variations in strength (softening), loadings (earthquakes) or groundwater pressures (high infiltration) must be considered
construction, monitoring and remedial measures	Establish setbacks for buildings and protect against toe erosion or removal of toe materials. Displacement or pore-water pressure monitoring as applicable (recommended if $F < 1.3$). Slope flattening, slope strengthening where applicable, slope drainage when material has a high frictional strength component and r_u is high

possibility of residual strengths developing in these layers. Strength anisotropy can be investigated by testing samples orientated in different directions (see, for example, Lo 1965). Low effective normal stress exists in slopes, and the effective stresses are decreased by increased seasonal water pressures. Under these stress conditions long-term softening of the soil mass can develop or dilative failure can develop in closely fissured soils or highly fractured rocks (see Ch. 2). It is the responsibility of the earth structures engineer to ensure that the failure mechanisms and test data used in estimating safety factors are obtained by correct procedures. Table 4.3 summarizes the basic procedures and considerations presented in the preceding sections and Table 4.4 lists some case studies involving slope failures in different types of earth materials.

4.6   Construction considerations and remedial measures

Figure 4.23a shows what can happen when rigid storm drain pipes are placed in slopes and Figure 4.23c shows a landslide initiated when a road fill was placed on sloping terrain. Figure 4.23b shows approach cuts at a bridge crossing (under construction) to reduce the risk of landsliding in an earthflow-prone area.

Services, particularly gas or water mains, should not be placed in slopes where creep movements may develop, causing rupture of these services. Stability should be considered before service trenches, irrigation, underground water disposal or construction activities are carried out on upland terraces. Removal of toe support and crest loadings should be avoided unless detailed analysis of the effects are carried out. Where major construction projects are carried out close to natural slopes, slope monitoring

Table 4.4   Some case studies of slope stability problems.

Case studies	Brief description	Reference
Aberfan disaster (1966); Wales	Conical piles of loose partly saturated coal wastes became unstable due to saturation by springs and rainfall. This material collapsed and flowed nearly 1 km burying a school and over 100 children	Aberfan Tribunal Report (1968), Bishop (1973)
Turnagain Heights landslide (1964); Alaska	A classical example of earthquake-induced liquefaction and flow of saturated sands. The liquefaction phenomenon and the flow properties of liquefied sand are discussed	Seed and Wilson (1967)
earthflows in sensitive clays; Canada, USA and Scandinavian countries	Sensitive fine-grained marine sediments flow laterally from a flat terrace into adjacent water or valley. Rotational retrogression, lateral spreading and possible liquefaction of silt and sand layers are causes	Tavenas et al. (1971), Bjerrum et al. (1971), Mitchell and Markell (1974)
massive rockslides; Canada, USA Europe	Massive rockslides have developed in mountainous areas where folded or blocky rock formations develop a fairly continuous sliding plane that closely parallels the slope face	Cruden and Krahn (1973), Muller (1964b)
softening and residual sliding in overconsolidated clays and clay shales	Heavily overconsolidated materials are subjected to increasing shear stresses beneath an excavated slope since the horizontal stresses do not decrease in direct proportion to the removal of overburden stress. This can cause a generalized strain softening or can result in movements being concentrated in weak planes. The result is long-term strength decrease and creep movements or failure	Skempton (1954, 1970), Bjerrum (1967b), Deere and Patton (1971)

145

Figure 4.23 Construction considerations. (a) Slope movements can pull pipes apart and services should not be directed downdip in a slope. This uplands drainage pipe concentrated ground water and caused landsliding and erosion in the slope. (b) Approach cuts and granular fill around abutments will protect this bridge structure from being damaged by slope movements. (c) Roadway embankments placed on naturally sloping terrain can cause landsliding. Cuttings with adequate drainage or limited embankment heights should be used in many cases.

instrumentation (piezometer and inclinometer installations) should be included if there is any risk of slope movements.

Remedial measures can be grouped under three headings, geometrical improvements, slope drainage and slope strengthening, the appropriate measure(s) depending on the material properties, type of failure anticipated and present groundwater conditions. Remedial measures and slope monitoring are briefly discussed below and detailed case record studies have been published by Zaruba and Mencl (1969), Royster (1980) and Gedney and Weber (1978).

4.6.1 Geometrical improvements

Slope benching can produce marginal improvements in the safety factor if the benching produces a reduced overall r_u factor and the material has a frictional strength property. Major improvement in safety factor is achieved by slope grading (flattening) or construction of a toe berm. The latter is particularly effective where toe erosion has produced an adverse effect. Figure 4.24b shows a rock berm in place with the minimum weight of armor stone related to the wave height (typically about $0.01h^3$ tonnes, where h is the wave height in meters) and the interior of the berm composed of a relatively pervious material which will not be subjected to internal erosion. Synthetic filter fabrics or graded filters can be used to prevent internal erosion of piping. Gabions (rock pieces in wire baskets) can also be used for erosion protection of stable slopes or as a replacement for armor stone. In all cases, durable rock types should be selected. Toe berms are most effective in materials exhibiting some frictional shearing resistance and least effective where underlying weak layers promote deeper failures.

In some instances the long-term effects of shoreline erosion and landsliding can be drastically reduced or eliminated using groins, breakwaters, beach filters or combinations of these to create a stable beach. References to successful applications of these methods may be found in coastal engineering literature.

4.6.2 Slope drainage

Gravity drainage of slopes is, in theory, an effective stabilizing method in a frictional material having significant groundwater pressures. In practice, excellent results have been obtained in coarse-grained soil and fractured rock slopes, but the following problems are noted in fine-grained soils.

(a) Horizontal or inclined perforated pipe drains installed from the toe of the slope have a fairly localized effect and, while the critical arc may be stabilized, shallower or slightly deeper failures can develop. To be effective, pipe drains must be installed with a graded filter surround. This can be accomplished using hollow core augers.

147

Figure 4.24 Slope protection methods. (a) Toe of slope prior to remedial work. (b) A rockfill toe berm was placed to eliminate active toe erosion and stabilize the river bank at this location. River currents, wave action and iceflows are factors which cause erosion. (c) Shotcrete operation and shotcreted rock slope. Wire mesh or wire fibres are often used to provide tensile strength. (Photos in (c) courtesy of John D. Smith Engineering Associates Ltd.)

(b) In clay and clayey silt soils a smeared zone may be formed around the drainage pipe during installation which would reduce the effectiveness of the drain. Installation by wash-boring techniques is recommended to avoid this problem. Natural variations in permeability with depth can also reduce the effectiveness of lateral drains.
(c) Creep movements can damage drainage installations, rendering these ineffective after several years. Damaged pipes become clogged with sediment or filter materials.

Wick drains that can be inexpensively injected into soils to provide extensive coverage are useful where there is a pervious drainage layer at depth. Inclined drains have also proven effective in layered systems where these drains can be pushed into a relatively pervious layer (LaRochelle et al. 1977). Suction wells (well points) can be effective in temporary stabilization of most soils during the planning and construction of permanent remedial measures.

4.6.3 Slope strengthening

Rock bolting, shotcreting (with wire fibres or mesh) and concrete abutment walls are effective in strengthening low- to medium-height fractured rock slopes, particularly where surficial weathering is creating stability problems. Drainage holes must be provided through concrete facings to prevent water pressure build-up. Figure 4.24c shows a shotcrete operation and a shot-creted rock slope. Hoek and Bray (1974) provide detailed considerations with regard to bolting and bracing rock slopes, and Piteau and Peckover (1978) describe general cases where rock slopes have been successfully strengthened.

Gravity retaining walls are used at the toe of soil slopes to prevent erosion and to provide greater space for developments of roadways and other facilities. These abutment walls generally do not improve the safety factor of soil slopes because rotational failure arcs can pass below the base of the wall. When permanent excavations to a more competent strata are carried out, however, a gravity or cantilever wall founded in the more competent strata can improve the stability of a slope. Suitable drainage and protection against frost action must be incorporated as shown on Figure 4.25. The slope angle should be designed with a safety factor in excess of unity over the full height (H_1), and the wall should be designed to provide a resistance in excess of that provided by the toe material excavated, i.e.

$$P \geq \tfrac{1}{2}\, \gamma H \tan \phi' + c' H \cot \beta$$

CMFE (1975) recommends a design pressure of $\tfrac{1}{2}\gamma H \cot \beta / \tan(45° + \phi'/2)$ and provides design charts for toe retaining walls. It should be noted that this

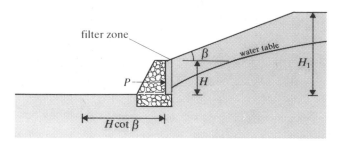

Figure 4.25 Slope retaining wall.

type of construction also requires consideration of the temporary factor of safety during construction.

It may be reasoned that strengthening members such as piles might be formed to provide additional shearing resistance in a plane of residual sliding but the scale of most residual landslides would make this proposition very expensive compared to alternatives such as drainage or chemical treatment. Soil nailing (using steel reinforcing bars pushed into an excavation face as excavation proceeds) has been used with success in granular soils with low r_u values (Gassler & Gudehus 1981).

4.6.4   Slope monitoring

Where slope stability is suspect or where the stabilizing effect of remedial measures requires monitoring, various systems have been developed to monitor slope movements or slope distress. Piezometer installations provide groundwater pressure data for analysis of current safety margins but do not provide a direct measure of slope distress. In open pit mining operations, pit slopes are often monitored using electronic distance-measuring equipment and photogrammetric techniques (Hoek & Bray 1974). Temporary monitoring of soil slopes is often carried out using slope stakes (wooden stakes driven into the sloped surface) and standard surveying equipment. Commercial inclinometer tubes and acoustic (or microseismic) devices are available for permanent monitoring of earth slopes (Koerner et al. 1978). These systems generally require detailed data interpretation by experienced engineers and are not readily adaptable as simple public-warning devices. Studies of slope movements show that significant surface movements generally develop well in advance of slope failures and this evidence favors the use of simple surface extensometer warning devices (see, for example, Mitchell & Williams 1981). Wilson and Mikkelsen (1978) provide detailed descriptions of many commercial slope-monitoring devices.

4.7 Permanent retaining walls

In some cases permanent earth anchors can be used to resist movements of foundations and structures (see, for example, Trow 1974) but earth retaining structures are usually of the gravity or cantilever type. Gravity walls are required to resist overturning by virtue of their self-weight and become uneconomical for all but modest heights. Reinforced concrete cantilever walls are generally more economical, since the weight of backfill materials contributes to the resistance to overturning as shown on Figure 4.26. Granular backfill materials are normally specified in order to provide drainage and eliminate frost action or shrinkage and swelling behind the wall. The total force on the wall shown on Figure 4.26 may be estimated as

$$P = (\tfrac{1}{2}H^2\gamma + qH)K_a - 2Kc'H \qquad (4.21)$$

where K_a may be taken as $\tan^2(45° - \phi'/2)$ and K varies from unity for mainly frictional materials to a value of 2 for mainly cohesive materials.

The force P is assumed to act at a depth of two-thirds of the wall height. In the cases of saturated backfills behind quay walls subjected to wave action or soils that may be subject to liquefaction during earthquakes, the design wall forces will be different from those approximated by Equation 4.21 (see, for example, Seed & Whitman 1970).

Considerations of bearing capacity, sliding, overturning and settlement must be examined in the design of a permanent retaining wall. The vertical stress beneath the cantilever base will vary approximately linearly across this

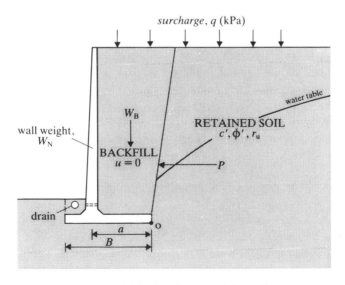

Figure 4.26 Cantilever retaining wall.

151

base and may be calculated as the material weights per unit wall length divided by the base width and multiplied by an eccentricity factor. Thus

$$\sigma_v = \frac{W_w + W_B}{B} \left(1 \pm \frac{6e}{B}\right) \tag{4.22}$$

where the eccentricity of the resultant force is determined by dividing the sum of the moments of all forces about the inside corner of the base (point O) by the total vertical force and subtracting half the base width. Thus

$$e = \frac{PH/3 + a(W_w + W_B/2)}{W_w + W_B} - \frac{B}{2} \tag{4.23}$$

From Equation 4.22 it is noted that the value of e must be less than $B/6$ or uplift forces will act on the base of the wall. If water pressures act at the base, the uplift force due to these pressures must also be included in the analysis. Provided that the base pressures are everywhere positive and the bearing capacity of the soil is not exceeded, the wall is safe from overturning. The factor of safety against lateral sliding of the wall is given by

$$F = (W_w + W_s)\mu'/P \tag{4.24}$$

where μ' is the coefficient of friction between the base and the wall. The wall base should be cast with sufficient roughness so that the friction will approach $\tan \phi'$ in a granular soil. A stiff cohesive soil has a low frictional resistance, and the factor of safety may be best evaluated as

$$F = CuB/P$$

where Cu is the undrained strength of the subgrade prior to construction.

Settlements under permanent retaining walls founded in compressible subgrades can result in tilting and stress concentrations in the wall or base, particularly when the eccentricity is not kept to a minimum. When retaining walls are designed to be rigid such that little or no lateral yield (bending) develops as the backfill is compacted, the lateral force may approach a value equal to

$$P = K_0 \gamma_B H^2/2$$

where K_0 is the lateral stress coefficient in the compacted backfill. It should be ascertained that the wall safety factors are sufficient to provide for this possible stress condition.

Vertical rock cuts and select compacted granular backfill can be stabilized by reinforcing members and adequate drainage. In such cases a facing wall

152

of light structural materials is often provided to prevent direct weathering of the slope face. Figure 4.27 shows an example of a gently dipping sedimentary rock outcrop stabilized by drainage and rock bolting and protected by a galvanized steel facing. Reinforced Earth® is a registered trademark of The Reinforced Earth Company, and is a proprietary design whereby metal reinforcing strips are placed in a select backfill material during placement and compaction, as shown on Figure 4.28. These strips prevent lateral deformation in the compacted soil mass by developing frictional resistance between the soil and the reinforcing strips. The lateral earth pressure on the facing segments is then negligible and drainage is provided to eliminate water pressures. A variety of architectural facings are available to provide esthetic appeal as well as to prevent erosion. Within practical limits, the height of a Reinforced Earth structure appears to be limited only by the horizontal space available for placement of the lateral reinforcing strips. The principles and mechanics of Reinforced Earth were introduced by Vidal (1966, 1969) and further detailed analyses of Reinforced Earth under static and dynamic loadings are contained in Lee *et al.* (1973), Kennedy *et al.* (1980) and Richardson and Lee (1975).

Some dense silts and sands are naturally reinforced (loess, oil sands) by particle interlocking due to solution weathering, or by cementation (see, for

Figure 4.27   Galvanized metal facing on bolted rock slope.

153

Figure 4.28 Reinforced Earth® structure under construction. (Photo courtesy of The Reinforced Earth Company.)

example, Dusseault & Morgenstern 1978) and high vertical slopes can be cut in these materials. Sand backfills can also be reinforced by cementation (underground mine-tailings backfills, constructed with hydraulically poured cement, have been exposed to vertical heights of 50 m with less than 5% normal Portland cement by dry weight). When a select backfill material is to be used in a retaining wall installation, methods of reinforcing this backfill may provide a more economical design than either cantilever or gravity walls.

4.8   Problems on slope stability

The design of major slope cuttings involves economic analysis as well as technical analysis. Capital costs associated with obtaining a larger safety factor (additional right-of-way acquisition and additional excavation work) may be balanced against costs of slope monitoring and remedial or repair measures. In order to estimate costs of remedial work *a priori*, the risk of slope failures must be determined. Statistical variations in design parameters, field correlations and even large-scale field testing may be required (see, for example, Kwan 1970) to provide data for risk analysis. In some cases of open-cast mining the slope angle is restricted by dragline dimensions, and this mining method is only economical if the slope geometry can be maintained within specified limits. The example problem below involves some of these considerations.

4.8.1 Example problem of the design of a shipping canal

A shipping canal 14 km in length, having a depth of 13.5 m (water depth 9.1 m) and a bottom width of 70 m, is to be constructed in an inactive homogeneous and lightly overconsolidated ($OCR = 2.5$) silty clay. The water table is at 2.5 m depth and the clay is underlain at 20 m depth by a stiff but relatively pervious silty sand over bedrock at 30 m depth. Average properties in the silty clay layer are:

$$C_c = 0.45$$
$$Cu = 40 + 2z \text{ kPa (where } z \text{ is in meters)}$$
$$\rho = 1770 \text{ kg m}^{-3}$$
$$c' = 12 \text{ kPa}$$
$$\phi' = 22.5°$$
$$w_L = 58\%$$
$$PI = 25\%$$
$$LI = 0.55$$

The values of c' and ϕ' were obtained at low normal effective stresses ($\phi'_N <$ 150 kPa) and the strength envelope was found to be linear with no sample size, anisotropy or work-softening effects in this range. From considerations of variations in strength and established field confidence in the rotational failure analysis, it has been estimated that the probability of failure of excavated slopes standing for two years is given by $P = 0.8/F^7$ where F is obtained using Equation 4.5 with drained triaxial test data and an average r_u value. The estimated construction time is 14 months, but due to winter shutdown the completion date is estimated to be 20 months after starting. The cost of excavation and haulage is estimated to be \$2/m³ if done by conventional earth-moving equipment or \$3/m³ if done by a shovel operation. This cost difference results because the excavated material must be hauled 1.5 km on average and compacted to form drumlin-shaped hills for a recreational area. There are two basic methods of construction:

(a) Use short-term slopes designed to be stable for a certain design period, at which time the water is brought up to operating level. Temporary dikes would be left in place between each excavated section and would be removed after the subsequent section was filled.

(b) Use long-term slopes designed to a suitable safety factor (with or without de-watering) and inundate the canal after completion.

The major cost to be considered is the cost of excavation and haulage. Because of the high water table and relatively high natural water content and plasticity of the soil, excavation by conventional scrapers would be limited to material above the water table. Temporary de-watering to lower the water table would, however, involve settlement of adjacent ground and potential damage to structures founded close to the canal. The pervious underlying silty sand could be used as a gravity drainage layer by constructing lines of wells into this layer on each side of the canal excavation, and creating sufficient drawdown to lower the water table below the canal bottom. Recharge wells could be used to limit the zone of influence and reduce consolidation settlements, but these considerations are beyond the scope of the current problem. For detailed discussions of canal de-watering considerations in a similar situation, the reader is referred to Farvolden and Nunan (1970) and Frind (1970).

In order to consider excavation costs the design slopes in (a) and (b) above should be compared to the ultimate canal design slope. If it is initially assumed that the canal slopes may be subjected to full rapid drawdown and a safety factor of unity is used to analyze this condition, then

$$\tan \phi'/F = 0.414$$

$$c'/\gamma HF = 0.051$$

and

$$c'/\gamma H \tan \phi' = 0.124$$

Since $\gamma_w/\gamma = 0.57$ is sufficiently close to $r_u = 0.6$, Figure 4.17 can be used without further iteration to give $\cot \beta = 2.9$ (a 3:1 slope would be selected). For a fully submerged 3:1 slope the safety factor is obtained, using $c'/\gamma' H \tan \phi' = 0.28$ in Figure 4.15, as $F = 2.58$. With the actual 9.1 m of submergence, $L/H = 0.33$ and the operating factor of safety of a 3:1 slope would be estimated with reference to Figure 4.22 as 1.88. If full rapid drawdown is not intended, the same design procedures may be followed to show that a 2.5:1 slope would provide an operating factor of safety of about 1.6 while a 2:1 slope provides an operating safety factor of about 1.5. While a canal slope would be subjected to some marginal drawdown in the wake of passing ships it would not normally be designed for full rapid drawdown. Slope protection for the effects of wave action would be provided and a design slope angle of 2:1 would appear to be appropriate.

156

For temporary excavations with $\Delta Cu/\Delta zy > 0.08$, Equation 4.19 is used in conjunction with Table 4.1 to give $F = 6.3(0.171) + 6.4(0.115) = 1.8$ for a 2:1 slope. Using Equation 4.20 for silty clay, this slope would stand for about 2.5 months. This would allow temporary excavation in sections of 1 to 1.5 km allowing time to fill the excavated sections. Shovel excavation would be required. For long-term excavation slopes using a safety factor of 1.3 and $c'/\gamma H \tan \phi' = 0.124$, $c'/\gamma HF = 0.039$ and $\tan \phi'/F = 0.318$, the following values of slope angles are obtained:

$$\text{for } r_u = 0, \quad \text{Figure 4.15, } \cot \beta = 1.7$$
$$\text{for } r_u = 0.3, \text{ Figure 4.16, } \cot \beta = 2.3$$
$$\text{for } r_u = 0.6, \text{ Figure 4.17, } \cot \beta = 4.3$$

The existing groundwater table would result in an average r_u value in the order of $0.95\gamma_w/\gamma = 0.54$ (see Fig. 4.13) and a slope of $\cot \beta = 3.9$. The difference in cost between slopes excavated at 2:1 and 3.9:1 is given as the cost per cubic meter times $(H^2/2)(\cot \beta_2 - \cot \beta_1)$ or about \$440 per meter of canal length for power shovel excavation. With major de-watering $(r_u = 0)$ the operational design slope of 2:1 could be excavated at a safety factor of about 1.5 using conventional earth-moving equipment with a very low risk of any slope failures $(P \simeq 2\%)$. Unless major de-watering is likely to cause considerable settlement damage in the vicinity of the canal, this method of construction would be recommended under the conditions given.

4.8.2 Tutorial problems

Problem 1 Using upper-bound analyses, obtain a safety factor for the graben formation mechanism shown on Figure 4.29a. Compare your answer to Equation 4.15 for the case of $c' = 0$, $h_w = 0$.

Problem 2 Using hand calculations and Equation 4.6, obtain a safety factor for the two failure arcs shown on Figure 4.29b and show that the deeper arc is more critical. Compare your safety factors to those given from Table 4.1.

Problem 3 An excavation to 12 m depth is to be opened in the soil profile characterized on Figure 4.18. Assuming \$2/m³ for soil slope excavation and replacement and \$309 per meter of perimeter for the installation of a vertical retaining wall, find the excavation durations when each method (temporary slopes or retaining walls) would be most economical. Assume low risk and adequate testing such that $F = 1.3 + 0.2M$.

Problem 4 Shear box tests on a partly saturated compacted sandy soil at $\gamma = 18$ kN m⁻³ gave strength parameters $c = 9$ kPa and $\phi = 30°$ over a normal

(a) Graben formation

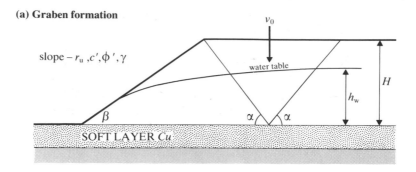

slope – r_u ,c',ϕ',γ

(b) Depth factor, d/H

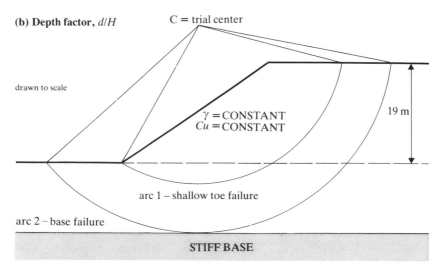

C = trial center

drawn to scale

γ = CONSTANT
Cu = CONSTANT

19 m

arc 1 – shallow toe failure

arc 2 – base failure

STIFF BASE

Figure 4.29 Slope stability problems.

effective stress range 5 kPa $\leq s_N \leq$ 100 kPa. This soil was compacted at γ = 18 kN m^{-3} to form a 10 m high impoundment dike. The design factor of safety of the downstream slope (using c' = 9 kPa, ϕ' = 30°) was F = 1.4 at r_u = γ_w/γ. This downstream slope failed within two months of water impoundment under static operating conditions. What was wrong with the design? Show calculations to support your answer.

Problem 5   A school building is located 100 m from a natural ravine slope with β_{avg} = 18.5°, H = 15m. The soil is a very sensitive clay (St = 16) with Cu_{avg} = 30 kPa, c' = 8 kPa, ϕ' = 32.5° (tan ϕ' = 0.637) and ρ = 1.7 t m^{-3}, overlain by 4 m of sand. Piezometer measurements at the building location indicate that the hydrostatic groundwater table is 7 m below ground level in the late fall and the high (springtime) water level is at the clay–sand contact. The area is in earthquake zone zero. Evaluate the stability of the natural

158

slope and the risk of the school being involved in a slope failure. If the risk is high, make remedial recommendations.

Problem 6　What is the most economical design slope angle for a very long uniform slope cutting of $H = 10$ m in a uniform soil with $c'/\gamma H = 0.025$, $\phi' = 25°$ and $r_u = 0.4$, if the probability of failure has been evaluated from statistical analyses as $P = 0.8/F^7$? The cost of slope excavation is \$2/m³ and the cost of repairing a failed section is estimated to be \$600 per meter of strike length.

Problem 7　The sketch on Figure 4.30a shows a typical section of a large-dimension stone quarry that has one quarry face parallel to and a distance D from a river. The quarried limestone is interbedded horizontally with thin, continuous and pervious layers of weak shale. In the long term the shale can be assigned a residual frictional shearing resistance given by

$$\tau_f = \sigma'_N \tan \phi'_r$$

An engineering report suggests that $D \geq 2H$ to ensure stability of the quarry wall.

(a) From the definition

$$F = \frac{\text{resisting forces}}{\text{driving forces}}$$

　　derive an accurate expression for F and note the assumptions which give the engineering approximation.
(b) Briefly discuss other potential mechanisms of failure of this quarry wall.

Problem 8　Typical natural slopes in oil sands tend to agree with the results of laboratory testing of undisturbed samples (Dusseault & Morgenstern 1978). Pit slopes excavated by dragline have stood at about 70° and 50 m height for the short-term mining sequence of several months. Natural slopes of this height tend to have average angles of about 55°. The oil sands are rather complex because they have a dense interlocked grain arrangement that is easily disturbed, internal gas pressures that depend on the ambient pressure and temperature, and various degrees of water wetting of particles prior to bituminous impregnation.

　　With reference to Figure 4.8, plot the field critical strength envelopes for a 50 m slope height using $r_u = 0$ for pit slopes and $r_u = 0.25$ for natural slopes. With reference to Figure 4.20, estimate the additional cohesive strength that would be required to support a dragline on the top of the pit slope (70° slope) if the total dragline weight of 50 MN is considered to be spread over a width of 5 m.

159

(a) Quarry wall stability

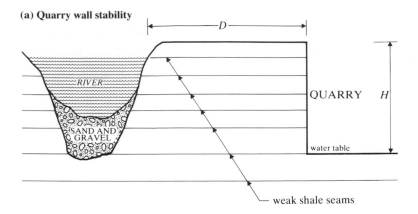

(b) Slope stability in varved sediments

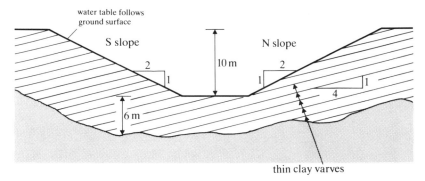

(c) Joint strength measurements

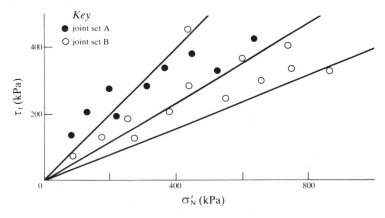

Figure 4.30 Problem profiles.

Problem 9 An east–west highway cut was made through a varved lacustrine deposit. The closely spaced alternate layers of silt and clay dip at 14°S as shown on Figure 4.30b. Two years after construction, during heavy rainfall, a slope failure occurred along the section.

(a) What mode (type) of failure would be likely to develop in the north slope?

(b) Assuming $c' = 0$, calculate the value of ϕ' giving $F = 1$ for non-circular sliding of the north slope.

(c) Assuming $\phi' = 0$, calculate the value of c' giving $F = 1$ for non-circular sliding of the north slope.

(d) Plot the two Mohr–Coulomb envelopes calculated in (b) and (c) above and show that they intersect in τ, σ'_N space at the average value of σ'_N acting on the failure surface.

(e) What mode (type) of failure is likely to develop in the south slope?

(f) Assuming $c' = 0$, calculate the value of ϕ' giving $F = 1$ for circular arc sliding of the south slope.

(g) Assuming $\phi' = 0$, calculate the value of c' giving $F = 1$ for circular arc sliding of the south slope.

(h) If the two Mohr–Coulomb envelopes calculated in (f) and (g) above were plotted as in part (d), would they intersect at a value of σ'_N equal to the average value along the critical circular arc failure surface? Explain!

(i) Which slope is likely to be more critical for a typical varved clay?

Problem 10 Figure 2.22 shows a flow net for a cut slope; assuming $h = 10$ m, $H = 15$ m, $\cot \beta = 3$, $c' = 8.5$ kPa, $\phi' = 30°$ and $\gamma = 18.3$ kN m^{-3}, estimate the factor of safety of this slope under static conditions and under earthquake accelerations corresponding to zone 2 on Figure 4.21.

Problem 11 Two major joint sets have been located in a rock mass through which a road cutting is planned. They are: Set 1, striking N60°W with NE dip of 65°; Set 2, N30°E with SE dip of 56°. The cutting will be 20 m maximum height. The joint strength properties have been estimated from testing small core samples and the data is plotted on Figure 4.30c.

(a) Using sketches as required, show how the data on Figure 4.30c may have been obtained (i.e. test arrangements and the method of obtaining τ_f, σ'_N from the test sample result).

(b) The road cutting will enter the rock mass heading due south and a 200 m radius curve will be cut through the rock such that the roadway will exit from the rock cutting heading due west. Specify the slope cutting angles and any other recommendations regarding the stability.

Problem 12 A railway cutting was made to rock at a depth of 9 m at $\cot \beta = 3$, in a residual soil having $c' = 6$ kPa and $\phi' = 28°$. The residual strength in the actively weathering contact was given as $\phi'_r = 15°$ and this contact dipped at $10°$ toward the railway cut. At the time of construction the water pressures were everywhere equal to zero in the soil but the degree of saturation was close to 100%. After completion of this railway line, a safety engineer refused to allow trains along this section of track on the third day of continuous rainfall. Can his position be justified by slope stability calculations?

Problem 13 Rotational failure is the most common mechanism of failure in homogeneous slopes. If the groundwater pressures can be adequately represented by an r_u factor and the slope profile is simple, stability can be adequately evaluated from a chart.

(a) Calculate and plot the variation in F versus $\cot \beta$ for $2 \leq \cot \beta \leq 5$, $c'/\gamma H$ $= 0.05$, $\tan \phi' = 0.5$ and $r_u = 0.6$, and comment on the form of the relation.

(b) Assuming that the factor of safety of a given slope varies linearly with r_u according to the equation $F = m - n(r_u)$, calculate and plot the slope ($n = \Delta F/\Delta r_u$) versus $\tan \phi'$ for $c'/\gamma H = 0.04$, $\cot \beta = 4$. (Use $\tan \phi'$ values of $0.2, 0.4, 0.6, 0.8$ and obtain F using only $\tan \phi'/F$ values.)

(c) Using $c'/\gamma H \tan \phi'$ values of $0.1, 0.2, 0.3, 0.4$ and obtaining F in the same manner, plot the variation in $\Delta F/\Delta r_u$ versus $c'/\gamma H$ for $\cot \beta = 4$ and $\tan \phi' = 0.2$.

(d) From the relations produced in (b) and (c) state a general conclusion with regard to the effect of slope water pressures on the stability of slopes in different types of materials, and note what general form of strength criterion is most amenable to stabilization by slope drainage methods.

Problem 14 A roadway embankment 2 m in average thickness (to resist frost heaving and provide adequate pavement design) is to be constructed at mid-height and along the strike of a long $5:1$ natural clay slope face which exists at a safety factor of 1.3. Discuss the following with the aid of sketches:

(a) What are the advantages in terms of safety of constructing a cut into which the embankment will be built as opposed to constructing the embankment on the slope face?

(b) What are the possible effects of the embankment on groundwater pressures in the slope? What local drainage facilities would be recommended in conjunction with the roadway construction?

Problem 15 Does the building shown on Figure 4.14 have a factor of safety of 1.6 or greater against being involved in a landslide? Provide arguments to support your answer.

5 Earth dams

5.1 Types of earth dams

Earth dikes and dams are gravity structures used to impound water or semi-fluid wastes. Water is impounded for flood control, hydroelectric power production, recreation, domestic storage, and industrial uses, including waste storage. Small dikes and dams are often homogeneous (constructed using one basic material) but most large dams (over 15 m) are composite or zoned dams and are constructed from two or more basic materials. Dam heights and volumes have been progressively increasing in the search for hydroelectric power, commonly exceeding 100 m after 1950 and approaching 300 m in recent years. Several dam complexes with compacted earth volumes in excess of 10^8 m^3 have been constructed, and the James Bay Project of Hydro-Quebec, Canada, is destined to be one of the largest heavy construction projects in the world. Typical components of a large earth dam are shown on Figure 5.1, and Figures 5.2 through 5 show various aspects of dam construction at the James Bay LaGrande Project.

The design of small dams is controlled to a large extent by government agencies that provide specifications for minimum side slopes, spillway and overflow requirements, and other details. Every large dam is an individual engineering accomplishment in terms of material selection, design and construction, the latter often requiring extensive river diversion facilities. For example, the 91 m high Llyn Brianne Dam completed in Wales in 1972 required a 5 m diameter diversion tunnel drilled and blasted 400 m through rock (Carlyle 1969).

For all earth materials, $c'/\gamma H$ approaches zero as H gets large and the shell components of high dams depend mainly on frictional strength for stability. Rockfill is often selected for the shell material and these dams are referred to as rockfill dams, but they must have an impervious core to impound water.

Mine-tailings dams differ from other dams with regard to material acquisition and placement and construction scheduling – they are normally constructed of mine wastes, particularly hydraulically delivered tailings and are constructed over the lifetime of the mine operation, 10 to 50 years being common.

5.2 Dam design considerations

The first design consideration must be with regard to the probable effects of proposed reservoir levels on regional groundwater flow and regional

163

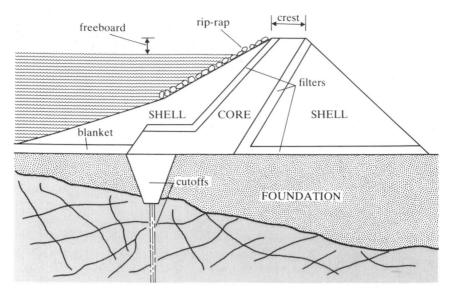

Figure 5.1 Typical components of a large earth dam.

Component name	Purpose of component
foundation	support dam and restrict underseepage
cutoffs	reduce seepage through foundation
blanket	alternative to cutoff for deep foundation
core	prevent leakage through the dam
filters	prevent particle migration and piping
shells	provide gravity fill and structural support
freeboard	prevent overtopping and provide storage
rip-rap	prevent wave and water erosion
crest	provide access and increase stability

stability. All available air photos, maps (topographic, bedrock, soils), well records, piezometric and climatic data should be studied and supplemented where required by preliminary site investigations. A reservoir will normally raise the surrounding groundwater levels and can cause flooding in adjacent lowlands, raising of well levels with the possibility of contamination, and saturation of slopes with the possibility of reservoir slope stability problems. In extreme cases there is the possibility that higher pressures in rocks may trigger seismic activity by reducing the effective normal stresses in fault zones. The failures at Vajont and Baldwin Hills reservoirs and Teton Dam (see Table 5.4) should remind engineers of the importance of reservoir location and water levels on stability.

The watertightness of the reservoir walls, particularly in the area of the proposed dam, is also of prime importance – evidence of fault zones, buried channels and the like should be looked for on maps and photos. Such

Figure 5.2 Blasting a spillway channel in abutment rocks. (Photo courtesy of Peter Kiewit Sons Co. Ltd.)

Figure 5.3 Cleaning the rock abutments. (Photo courtesy of Peter Kiewit Sons Co. Ltd.)

Figure 5.4 Dam base placement and compaction. (a) Placing contact till on prepared foundation. (b) Core and shell construction in progress. (Photos courtesy of Peter Kiewit Sons Co. Ltd.)

Figure 5.5 Placing rip-rap and armor stone. (a) Placing graded rip-rap on upstream slope. (b) Setting individual pieces of armor stone. (Photos courtesy of Peter Kiewit Sons Co. Ltd.)

features may require treatment by grouting. Special considerations with regard to permafrost distribution are necessary in northern regions (see, for example, Piteau 1972), and environmental problems due to impoundment or changes in river flows are generally more troublesome in regions with sensitive ecosystems. If no serious regional problems exist, this preliminary terrain analysis should lead to the selection of one or more potential dam sites for detailed investigation. Cost–benefit considerations control the exact location of a dam once it is established that construction is feasible. A natural constriction between two stable abutment walls with good foundation conditions and in close proximity to available construction materials is ideal, provided that natural flows can be economically handled during construction. Many earth dams have concrete spillways or intakes associated with them and construction can proceed, on a rock foundation, by diverting natural flow to one side while the spillway structure is built and then using the spillway to accommodate flow while the earth dam section is constructed.

The selection of an earth dam over a gravity or arched concrete structure is generally made on economic grounds, although weak abutment formations or inadequate foundation conditions can often preclude the construction of a concrete dam. Air photos are valuable in searching for earth dam building materials and this initial search is followed by sampling and laboratory testing (identification, grain size, Atterberg limits, water contents, compaction, compression, permeability and strength tests) to evaluate the suitability of these materials (see Tables 2.1–4). Swelling minerals and materials susceptible to leaching should be avoided in the selection of earth dam construction materials.

Basic design requirements for earth dams may be listed, following the US Corps of Engineers (USCE 1971), as:

(a) Embankment slopes must be stable under all construction and operating conditions, including reservoir drawdown.
(b) The embankment must not impose excessive stresses on the foundation or abutments.
(c) Seepage flow through the embankment, foundation and abutments must be controlled so that piping, sloughing, or removal of material by solution does not occur. Seepage flow quantities may also be limited by storage considerations.
(d) Spillways, outlet capacities and freeboard must be sufficient to prevent overtopping. Freeboard must include allowances for post-construction embankment and foundation settlements.

5.2.1 Embankment slopes and dam shells

The shell of a dam provides structural support for the core and is normally compacted to a fairly high density (95–100% of standard Proctor) at low moisture contents. Moisture contents dry of optimum are usually specified so that excess water pressures do not build up as the dam height is increased. Dam shells are designed to form stable slopes and circular arc analysis, with appropriate r_u values for construction, normal operation and drawdown, is generally used to determine relevant safety factors.

Depending on design conditions and filter arrangements, the upstream slope may be the same as or shallower than the downstream slope. To appreciate the purposes of flatter upstream slopes one might consider the

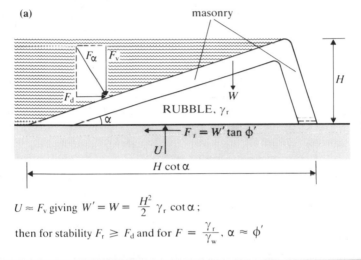

$U \approx F_v$ giving $W' = W = \dfrac{H^2}{2} \gamma_r \cot\alpha$;

then for stability $F_r \geq F_d$ and for $F = \dfrac{\gamma_r}{\gamma_w}$, $\alpha \approx \phi'$

Figure 5.6 Masonry dam stability. (a) Schematic diagram. (b) Masonry dam at Horseshoe Valley in Wales.

stability of older masonry dams such as shown on Figure 5.6. Reduced slope angles also increase the resistance to seepage flow through and beneath homogeneous dams.

Since most of the dam volume is comprised of shell materials, the availability and handling of shell materials is a major economic consideration. Drumlinized tills which can be readily excavated and compacted to stable slopes of 3.5:1 may be more economical for limited dam volumes than specially processed crushed rock with slopes of 2:1. In many large earth dams material savings have been realized by using toe berms, or by changing the shell slope at one or more elevations as shown on Figure 5.1.

Safety factors are used to express the stability of embankment slopes, and the slope stability analytical methods discussed in Chapter 4 are applied to obtain numerical values for safety factors. Recommended safety factors are included in Table 4.2. Although compacted materials are more homogeneous than natural formations, stress conditions and pore-water conditions in earth dams may not be homogeneous either during construction or during operation. Laboratory stress–strain relations, field stress measurements and field pore-water pressure measurements should be incorporated into calculations, particularly when a high risk is involved. Higher factors of safety should be used when confidence in the design parameters or in the construction control is not high and where weak foundation materials are present. Where weak foundation materials are combined with a weak core material, failure mechanisms such as those shown on Figure 5.7 should be considered. Stability analysis of sloping cores is discussed further by Seed and Sultan (1967).

5.2.2 Foundation design considerations

The foundation must support the dam and control underseepage – the two functions are quite separate; a fractured rock would provide adequate support but would not be watertight and would contribute to uplift pressures on the dam base. Bearing capacity, watertightness, settlements and foundation preparation are all important considerations.

Most rocks and cohesionless soils (sands and gravels) are excellent bearing materials. For cohesionless soils the safety factor of embankments can be estimated as

$$F = N_\gamma B/4H$$

where B and H are the base width and height, respectively, of the embankment. N_γ is related to ϕ' and varies from about 10 to 100 for granular materials of different densities. Since B must be greater than $2H$ for slope stability, F is usually greater than 5. The exceptions are loose saturated sands and silts under artesian pressures or which, under vibratory loading,

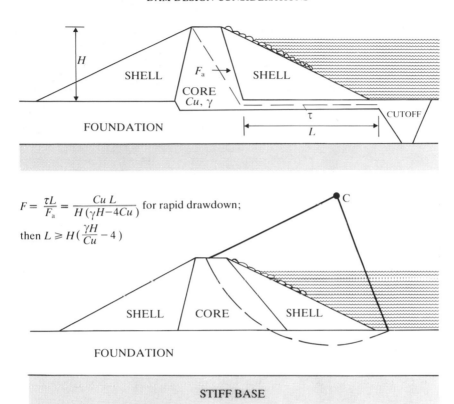

$$F = \frac{\tau L}{F_a} = \frac{Cu\,L}{H\,(\gamma H - 4Cu)} \text{ for rapid drawdown;}$$

then $L \geqslant H\left(\dfrac{\gamma H}{Cu} - 4\right)$

STIFF BASE

— · — · — circular arc through weak core and foundation analyzed by computer or by chart solutions

Figure 5.7 Failures due to weak cores and foundations.

may suddenly liquefy, and weak rocks such as Cretaceous shales, particularly if layers containing swelling minerals are present, which may exhibit softening and progressive lateral movements. Cohesive soils, on the other hand, are usually troublesome and often will support only modest embankment heights. Organic materials must always be removed and weak clays and silts should also be excavated and replaced when feasible.

The bearing capacity of a clay soil is normally estimated using

$$F = 5Cu/\gamma H$$

Thus, even a stiff clay with an average Cu of 100 kPa would have a limiting capacity of about 25 m of embankment height. When Cu is obtained from field vane tests in highly plastic soils, this strength should be checked by undrained triaxial testing or should be reduced by a vane correction factor as discussed in Chapter 3. Stabilizing berms can be used to extend the limiting height (an extra meter of height for each meter height of berm) but the berm

171

width must be substantial (see Section 3.1.1) and the shell material requirements may be increased by a factor of about 1.5. Low safety factors and the accompanying lateral spread of soft foundation soils should be avoided because cracking of the dam core could develop and this could lead to high leakage and piping failures. If a dam on a soft foundation soil is constructed over an extended period of time the foundation will have time to consolidate and strengthen, thus supporting additional load – this is referred to as stage construction. An approximate approach to calculating the bearing capacity of a soft soil during stage construction is outlined in Chapter 3. A preferred approach would be to use effective stress strength parameters c' and ϕ' in a circular arc stability analysis. In either case it is imperative that foundation pore-water pressures be monitored during stage construction to obtain groundwater pressures for use in the analysis.

With the exception of intact rock, which is the ideal dam foundation, good bearing materials are generally pervious while relatively impervious soils are not ideal bearing materials. A flow net can be constructed to check the underseepage of a proposed dam – if leakage is too high or critical flow gradients are expected to develop, a cutoff or other foundation treatment will be necessary. Foundation treatment methods are detailed in Section 5.3.

As with bearing capacity, it is the soft cohesive alluvial deposits, found in most river valleys, which are troublesome with regard to settlement. One known exception is discontinuous permafrost, such as the foundation under dikes at Kelsey generating station, Manitoba, which produced up to 3 m of settlement upon thawing (Johnston 1969). Large settlements can, of course, be tolerated provided that the integrity of the dam and cutoffs are preserved; settlement of the Manicouagan-3 Dam was calculated to be several meters during a 15 year period following its construction in 1975. It is the differential settlements, which can cause distortion and cracking within the dam, that must be avoided or designed for. Dam settlements and distortion calculations are discussed in Section 5.4.

5.2.3 Cores and cutoffs for seepage control

Dam cores are normally placed slightly upstream of the sectional centerline to form the impervious part of the dam. Cores and cutoffs, as previously noted, reduce the seepage and uplift. Storage dams may have prescribed leakage limits (typically 0.1% of the stream flow) while larger leakage could be tolerated in flood control dams provided that this leakage is controlled and does not lead to piping or other instabilities. Uplift pressures are limited by stability calculations, and it is usually recommended that large safety factors be applied to critical hydraulic gradient or exit gradient calculations in cases where underseepage is not intercepted by filters or relief wells.

Where suitable soil is available, an earth core will be cheaper, safer and more durable than alternatives such as concrete, steel sheeting or rubber sheeting. Sandy clays and clay tills of low to medium plasticity ($PI < 30$) and having a compacted permeability less than 10^{-7} m s^{-1} are ideal core materials. Compaction should be between 90 and 98% of standard Proctor depending on the material and the expected distortion; higher compaction for plastic soils or small settlements and lower compaction where larger settlements are expected or where less plastic materials are used. Core compaction can be controlled by controlling placement moisture content at a few percent above optimum water content in order to produce a more homogeneous and less brittle material. Construction pore-water pressures in the core may, however, be a problem when the core material is compacted wet of optimum. For structural purposes and to facilitate construction, earth cores are generally thicker than required for seepage control – typically the core thickness is about 5 m at the dam crest and increases with depth by about 0.1 to 0.3 m per meter.

Cutoffs should be located directly beneath the core or upstream of the core and joined to it by an impervious blanket. In positioning cutoffs and cores the designer must be careful that a weak link is not inserted in the dam in terms of leakage as well as stability. Cutoffs not only reduce underseepage but also reduce uplift on the downstream side of the dam. A cutoff may be partial such as core type material compacted into an excavated trench (see Fig. 5.8) or may extend to great depths to cut off pervious layers in the foundation. The Hydro-Quebec 107 m high Manicouagan-3 Dam has a double-wall cutoff composed of cast-in-place concrete interlocking piles and panels, extending through 126 m of pervious alluvial valley infill (Dascal 1979). Slurry trench walls and grout curtains are also effective cutoffs – the former being more applicable to easily excavated materials and the latter to fissured rocks and coarse gravels. If readily compacted low-plasticity clay

Figure 5.8 Core trench construction for a small dam. (a) Cutoff trench with hydraulic settlement tube installed. (b) Compacting impervious material in cutoff trench. (Photos by J. Agar.)

(a) Piping due to seepage beneath a weir

exit gradient i_e approaches the critical gradient i_c

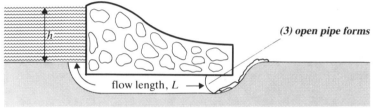

exit gradient increases due to erosion

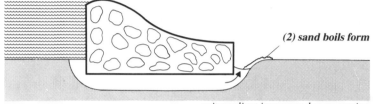

pipe regresses as h/L increases and reservoir is emptied

(b) Piping due to hydraulic fracture in core or cutoff of an earth dam

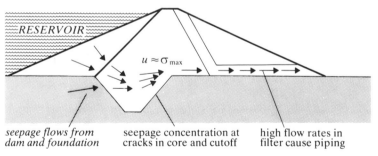

Figure 5.9 Piping failures.

material is available in sufficient quantity, an upstream blanket might be considered – a minimum thickness of 1 m with an increase of 1 m for every 10 m of water is recommended. The extent of the blanket would be determined from seepage considerations.

In all cases, cutoffs and blankets must be structurally integrated with the core of the dam to provide a continuous water barrier. The cutoff should be located upstream of the dam centerline to reduce uplift. Natural valley walls (abutments) must also be considered with respect to watertightness – cutoffs, cores and blankets should be made integral with watertight valley walls. Further details are provided in Section 5.3.

5.2.4 Filters and internal drains

Seepage water must exit from the dam and foundation. Exit gradients are of primary concern since these can cause internal erosion and piping. Piping is the term used to describe the rapid internal erosion of soil along a flow channel once exit gradients begin dislodging particles, and examples of this process arc shown on Figure 5.9. A chimney filter will help to prevent piping provided that no cracks develop in the core or near the top of the cutoff trench. This type of filter is designed to conduct all core seepage out of the dam shell, as shown on Figure 5.10. By removing seepage water the filter will also stabilize the downstream shell by reducing r_u. A base filter is compacted as a single horizontal sand layer about 0.5 to 1.5 m thick, and an inclined extension of the base filter can also be easily compacted onto a downstream starter dike to pick up slope seepage in a homogeneous dam. In a zoned dam a sand filter must be placed on the downstream side of the core to collect the seepage (see Fig. 5.1). If more than one material is necessary in order to prevent migration of soil particles, the filter is called a transition filter. Filter

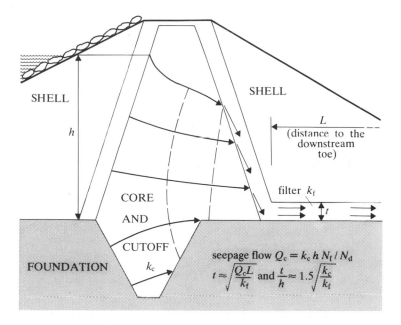

Figure 5.10 Flow through base filter.

materials should be well graded, contain less than 5% minus 200 sieve size and satisfy the following grading characteristics:

$$\frac{D_{15}(\text{filter})}{D_{85}(\text{soil})} \leq 5 \qquad \frac{D_{15}(\text{filter})}{D_{15}(\text{soil})} \geq 5 \qquad \frac{D_{50}(\text{filter})}{D_{50}(\text{soil})} \leq 25 \qquad (5.1)$$

A transition filter is necessary when the shell material does not fulfill the grading requirements of a filter with respect to the core filter. In a transition filter each subsequent material acts as a filter for the adjacent material. To transition from a fine-grained core to a rockfill shell would, for example, require at least two filter materials, the inner material being a well graded sand and the outer layer being a well graded gravel or crusher run material. Geotextiles are available under a variety of trade names for use in preventing soil particle migration at material interfaces. These should not be used in place of sand or gravel filters in earth dams, but may be used in conjunction with a clean sand or gravel which may not satisfy one of the grading requirements (the clean granular material being capable of conducting the seepage water to a drain and the filter material preventing particle migration). Internal drains should be avoided in earth dams whenever possible but, if required, can be constructed of fabric-wrapped perforated pipe located in a filter zone. These drains should be designed to conduct seepage water out of the dam by gravity flow. Cedergren (1973) provides some excellent examples of seepage flow nets in various configurations of earth dams.

Filters should be placed on both sides of the core and may be required on the downstream sides of compacted cutoffs or impervious blankets. Theoretically, the required thickness of a filter can be calculated from flow considerations (as noted on Fig. 5.10) to provide for the conduction of water out of the dam under small gradients. In practice, filters tend to be 1 to 2 m in thickness to facilitate placement (see Fig. 5.11). Thicker filters are often desirable in reducing the distortion due to differential settlements between cores and shells.

5.2.5 *Erosion protection and other considerations*

Rip-rap, extending over the entire freeboard and to a depth of about twice the aniticipated wavelength below the reservoir level, is used to protect the upstream face of an earth dam from wave erosion. Charts for wave forecasting are available in coastal engineering texts and handbooks, but if it is assumed that the wind duration will be sufficiently long to produce an equilibrium wave height and that the slope run-up will be 1.5 times the wave height, the minimum freeboard can be estimated as

$$\text{freeboard (m)} = 0.016UF^{0.4}$$

176

Figure 5.11 Placing filter zone in a homogeneous embankment dam. (a) Placing graded filter material to form inclined filter. (b) Placing and compacting embankment material. (Photos by J. Agar.)

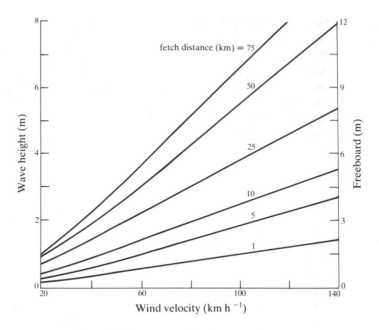

Figure 5.12 Freeboard for long-duration waves.

where U is the wind velocity (in km h^{-1}) and F the fetch distance across reservoir (in km).

Figure 5.12, based on equations presented by Sverdrup and Munk (1946) may also be used for freeboard design. Additional freeboard is required for storage and anticipated settlements.

Rip-rap is composed of good-quality angular broken rock, dumped or placed in two layers over a sand and gravel filter cushion. This filter cushion must prevent scour and undercutting of the rip-rap and could contain wire mesh and filter fabric. The wire mesh also prevents animal burrows. The outer layer of rip-rap is formed of the coarser armor stone with individual blocks being placed, as shown on Figure 5.5, to form a continuous interlocking surface. Hudson (1954) presents detailed calculations of armor stone sizing but the stone mass requirement can be estimated from the formula

$$M = 0.01h^3 \text{ tonnes}$$

where h is the wave height (in meters) as obtained from Figure 5.12. For a wave height of 5 m, for example, armor stone having a mass of 1.25 tonne (0.4 m^3 volume or 0.75 m on a side) would be recommended.

Downstream slope protection in the form of sodding or crushed rock over

Figure 5.13 Eastmain Dam and spillway structure at James Bay Project. (Photo courtesy of Peter Kiewit Sons Co. Ltd.)

a filter media is required to prevent wind or runoff erosion and limit the infiltration seepage into the shell. Crest widths are normally greater than 5 m to allow access for placement of rip-rap and routine inspection.

Spillways and overflows designed to pass 10 000 year return interval flood flows should be built into the abutments whenever possible. An outlet tunnel or a bypass channel (as shown on Fig. 5.13) is normally constructed for large dams. If the foundation is sufficiently competent, a central spillway with earth wrap-around of the concrete spillway or with concrete retaining structures at the spillway location is suitable for small dams (Sowers 1962). Conduits through earth dams should only be considered if the dam is homogeneous and the foundation is rigid. All such appurtenances should have cutoffs (flanges extending laterally from the conduit) to prevent piping at the conduit–earth dam contact and should be designed to withstand anticipated movements (see, for example, Rutledge & Gould 1973).

179

Emergency spillways may be located at natural saddles or saddle dams remote from the main dam and small dams may be designed to withstand emergency overtopping (Pravdivets & Slissky 1981, Middlebrooks 1977).

Reservoir leakage can often be greater than the leakage through or beneath an earth dam. Fault zones and buried channels may require grouting or the placement of an impervious blanket in the reservoir. Stability of the reservoir slopes under the proposed filling and operating schedules must also be considered and procedures for analysis have been outlined in Chapter 4. Reservoir slope stability problems are encountered when the water content of previously unsaturated materials is increased, particularly if swelling or easily weathered minerals are present. Slope slumping or failure can occur as the water level is raised. Erosion protection may be required where the reservoir level encounters easily erodible soils. Reservoir clearing and grubbing should be carried out to the high water level to remove all debris in the impoundment area. The effects of impoundment and operational flows on erosion and stability, sediment transport and other environmental considerations (both upstream and downstream of the reservoir) should also be considered.

5.3 Foundation treatments and efficiencies

Seepage control using cutoffs was briefly introduced in Section 5.2.3. Cutoffs are designed to create a flow barrier over which a large head loss will occur. The hydraulic efficiency of these types of flow barriers has been defined by Terzaghi and Peck (1967) as

$$E_{\mathrm{H}} = \frac{\text{head loss across barrier}}{\text{total head loss}} = \frac{\Delta h}{h} \tag{5.2}$$

Piezometer strings are installed on each side of the cutoff to monitor its efficiency, and efficiencies exceeding 90% have been attained (see, for example, Dascal 1979).

Treatment of the foundation material by zone grouting or densification is often suitable for moderate impoundment heights. These methods are sometimes used in conjunction with a partial cutoff and are not intended to create a water barrier. In such cases the effectiveness of foundation treatment has been defined by Casagrande (1940) as

$$E_Q = 1 - Q/Q_0 \tag{5.3}$$

where Q is the quantity of flow through the treated foundation and Q_0 is the theoretical quantity of flow without foundation treatment. Flow quantities are often difficult to determine accurately, but may be estimated from

Table 5.1 Cutoffs for dam foundations.

Cutoff type	Brief description	Suitable soil type and depths	Comments on performance	References
sheet piling diaphragm wall	drive plane or zeta piling with vibratory hammer and water jets; moderate costs	alluvium, no large boulders, to 30 m	efficiency often low unless upstream side grouted with bentonite slurry	Lane and Wohlt (1961), Middlebrooks (1942)
grout curtains	several rows of 2 to 3 m spaced holes grouted at high pressure with cement soil grouts; medium to high costs	fractured rock and coarse-grained soils to 150 m; $k < 10^{-5}$ m s^{-1}	fair to good efficiency reduces k by a factor 50 to 100; also used in abutments to control leakage	Wafa and Labib (1967), Marsal and Resendiz (1971)
concrete piles and panels (also precast concrete walls)	excavate bentonite slurry stabilized slot between piles and insert panels with grouted joints or tremie concrete panels; medium costs	weak rocks, alluvium to depths of 50 m; excavate with Kelly bar	good efficiency but requires special construction control and design	Galbiatti (1963), Dascal (1979), Marsal and Resendiz (1971)
slurry trench	dragline excavation of slurry stabilized trench; displacement backfilled with sand gravel-clay bentonite (well graded with 50 mm max. size); low costs	alluvium or coarse-grained soils to 30 m or greater	good efficiency, flexible and compressible; possible loss of slurry if soils are coarse	Jones (1967)
impervious blanket	impervious soil compacted over reservoir bottom and connected to internal core or extended as upstream core on dam face; costs variable	very deep foundation soil of suitably low permeability and good grading characteristics	effective if abutment conditions are good; downstream filter drains or relief wells may be required	Casagrande (1969)
removal and replacement	excavate above or below water level at stable slopes; dry excavation allows inspection and treatment of underlying material; cost variable	soils which can be economically excavated to required depth with or without seepage control (20 m common)	good efficiency if replacement soil compacted in dry excavation; fair to good efficiency if placed under water	

Table 5.2 Treatment of foundation soils.

Treatment method	Brief description	Suitable soil type and effective depth	Comments on performance	References
densification by blasting	blast vibrations, liquefaction and settlement	loose saturated silts and sands to 20 m	70–80% relative density obtained	Dupont (1958)
compaction piles, sand piles	hollow pile vibrations densify soil; sand piles formed on removal	saturated or dry sand to 20 m	relatively high densities obtained	Wallways (1964), Janes (1973)
vibro-probes with replacement; vibroflotation with water jets on probe	vibratory probe densifies soil by displacement and compacts selected replacement material	cohesionless soils and sandy tills or alluvium to 30 m	relatively high density and good uniformity obtained	Baumann and Bauer (1974), Basore and Boitano (1969)
vibratory rollers (to 50 tonne)	roller mass compacts with aid of vibration	cohesionless soils to 3 m	high density obtained	Moorhouse and Baker (1968)
dynamic consolidation (to 40 tonne)	repeated dropping of large mass on soil surface	cohesionless soils and insensitive cohesive soils to 15 m	relatively high densities obtained	Menard and Broise (1975)
grouting with slurries or chemicals	injected from boreholes, with lime water (for expansive soils), chemical grouts	soils with groutability ratio $\dfrac{D_{15}(\text{soil})}{D_{85}(\text{grout})} > 25$ $k > 5 \times 10^{-6}\ \text{m s}^{-1}$ unlimited depths	significant reductions in permeability and compressibility obtained; high cost	ASCE (1978)
consolidation by preloading	embankment preload or constructed in stages	soft compressible clays and silty clays	improves bearing capacity by consolidation over extended time period, time reduced by sand or wick drains	Ladd et al. (1972)
displacement construction	granular or rockfill advanced to displace soil; excavation of forward mud wave	very soft clays, organic soils to 20 m; blasting may be required in fibrous peats	controlled failure to avoid entrapment of soft soils. Not recommended for impoundment structures	Weber (1962)

accurate measurement of discharge and stream flow quantities. Treatment of the foundation material may also be carried out in order to improve foundation strength or to reduce foundation settlements. Usually such treatment is also effective in reducing seepage flow. Pervious sand drains have, however, been used as a foundation treatment to improve the strength of the soft clay foundation at Selset Dam (Kennard & Kennard 1962) and to drain melt water from underlying permafrost at Kelsey generating station dikes (MacDonald *et al.* 1960).

5.3.1 Treatment methods

A summary of foundation treatment methods is presented on Tables 5.1 and 5.2. Whenever possible it is preferable to excavate unconsolidated materials and expose underlying competent formations for inspection and treatment. The upper layers of a rock formation are often weak and highly fractured due to weathering. Weathered rock should be removed from the foundation and abutment areas (see Figs. 5.2 & 3) and a core trench can be excavated to deeper intact rock as shown on Figure 5.14. Cavities are often found in exposed rock and these must be cleaned out and filled with dental concrete such that the rock foundation provides a clean sound surface upon which earth materials can be adequately compacted (see Fig. 5.15). Abutment slopes must be shaped and sloped to appropriate angles (usually less than 1 : 1) and grouting (cavity and fissure grouting or an extended grout curtain) is often necessary to ensure watertightness of the abutments. Figure 5.16 shows a rock abutment shaped prior to construction of an earth dam. Lateral shaping of the abutment in the core area is done to provide a tight contact when the reservoir water pressure exerts a downstream thrust on the dam. The first layers of earth fill (sometimes called contact till) should be placed in

Figure 5.14   Core trench preparation. (Photo courtesy of Peter Kiewit Sons Co. Ltd.)

183

Figure 5.15 Rock foundation preparation. (a) Unusually deep cavity in rock requires filling. (b) Rock foundation after cleaning and dental work. (Photos courtesy of Peter Kiewit Sons Co. Ltd.)

15 to 20 cm lifts and compacted with hand-operated vibratory rollers until a continuous level surface is available for motorized compaction. Contact till placement followed by compaction of 0.5 m layers using a 50 tonne pneumatic roller is shown on Figure 5.4.

When excavation of natural soil materials is not feasible, in-place treatment involves either improvement of soil masses or cutoffs. Loose cohesionless foundation materials must always be densified to decrease embankment settlements and reduce the potential for liquefaction failure in the event of an earthquake. Figure 5.17 shows vibroflotation equipment, and ordinary pile-driving equipment can be used for vibrodensification or the placement of vertical reinforcing columns. Dynamic consolidation (compaction) is carried out by dropping large masses as shown on Figure 5.18. Natural earth subgrades should be scarified so that the contact material can be compacted integral with the subgrade material. Figure 5.19 shows a clay till material as compacted by sheepsfoot roller.

Deep alluvial valley infill materials present the combined problems of compressible layers and pervious layers, and the underlying rock materials may contain faults, fissures and cavities. Rigid cutoff walls such as the concrete pile or concrete wall techniques noted on Table 5.1 can be employed in such soils, but a compressible grout or bentonite zone must be provided between the top of the cutoff wall and the compact dam materials in order that settlements can develop without inducing excess stresses in the dam or in the wall (Wilson & Squier 1969). Negative skin friction (Bozozuk 1972) can transfer stresses from compressible foundation soils to rigid cutoffs, and this stress transfer must be considered in the wall design. Bentonite grouting

Figure 5.16 Shaped and cleaned abutment. (Photo by N. Leedis.)

185

Figure 5.17   Vibroflotation equipment. (Photo courtesy of Peter Kiewit Sons Co. Ltd.)

can be effective in reducing skin friction. Slurry trench cutoffs are relatively flexible and grout curtains, of intermediate flexibility, have the advantage that they can be easily extended into the abutments to form a continuous curtain. Very detailed site investigation and laboratory testing must precede grout curtain installation so that grouting material types and quantities may be controlled. Simonds (1977) presents a detailed review of the grouting carried out to correct defects in adesitic formations comprising the foundations and abutments at the 220 m high Hoover Dam. Combinations of two or more cutoff types are sometimes used in order to take advantage of their relative merits and economies.

5.3.2   Slurry trenching and grouting

Figure 5.20 shows slurry trenching operations. Steel pipes are driven into the ground to divide the trench excavation into sections so that the bentonite

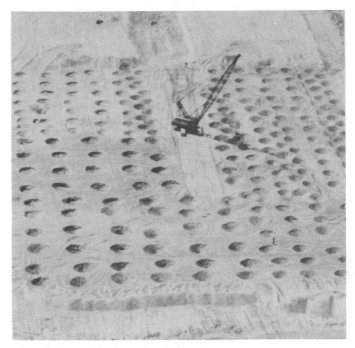

Figure 5.18   Dynamic consolidation. (Photo courtesy of Geopac Inc.)

Figure 5.19   Clay till dam compacted using a sheepsfoot roller. (Photo by N. Leedis.)

slurry used to support the trench walls can be re-used in subsequent sections. Excavation is carried out using a clamshell and an airlift pump for bottom cleaning. A thickened aggregate–bentonite slurry is then dumped into the trench to displace the bentonite slurry. The trench thickness, in this case, is close to 2 m but thinner slurry trenches have been used. Reinforced or plain concrete tremie walls may also be formed in a similar manner and the pipe piles can be used to provide interlocking of the wall sections.

Chemical gel grouts are available for use in material down to silt sizes. The grout is pumped as a liquid and gel set times from a few seconds to several hours can be arranged. The method has gained considerable success in recent years with sodium silicate and calcium chloride being mixed with acrylic monomers to form the grout. However, the method appears to be most applicable in marginal situations, requires careful control and may encounter environmental opposition. The chemicals and grouting procedures may often involve costs that would be considered excessive for earth dam foundation problems.

For coarser soils (coarse sands and gravels) cement and bentonite grouts are often used to form grout curtains. A typical configuration of a grout curtain is sketched on Figure 5.21 where two lines of grout holes, at 3 m centers, with 1.5 m offset should reduce the permeability of a typical sand and gravel to 10^{-7} m s^{-1} and a third line of holes should achieve a reduction to about 10^{-8} m s^{-1}.

A *tube-à-manchette* (tube with a sleeve) technique, as sketched on Figure 5.21, is commonly used to control grout injection. A tube is sealed into each borehole using a cement–bentonite grout which is allowed to set for seven days. A steel grout pipe with double packer assembly would then be inserted in the tube and injection grouting would be carried out, at each sleeve location, from the bottom to the top of the tube. A cement–bentonite injection grout is normally used at a pressure of 500 to 1000 kPa, but the pressure should be regulated to prevent ground heave and yet grout a sufficient volume of soil to reduce the permeability. A typical injection quantity is about 10% of the soil volume being grouted for compact sands, with the quantity increasing for loose uniform sands and gravels.

Grouting of sands and gravels decreases the permeability by compressing the material and intruding grout into the larger voids and the cavity formed by soil compression. Holes in the two-line pattern should be grouted in sequence whereas some advantage may be gained by delayed grouting of the central lines of holes in the three-line pattern. One advantage of the *tube-à-manchette* technique is that regrouting can be carried out, in whole or in part, without redrilling.

Careful records of water losses and soft drilling zones should be kept during drilling of grout holes in rock formations. These records provide data for selection of packer locations for grouting. Packers (mechanical or pneumatic) are used to isolate successive grout sections, from the bottom of the

Figure 5.20 Slurry trench construction. (a) Slurry trench excavation and cleaning with airlift pump. (b) Bentonite–granular slurry trench material placement. (Photos courtesy of Peter Kiewit Sons Co. Ltd.)

hole, as shown on Figure 5.22. Cement–water or cement–bentonite–water grouts can be used for fissures and small cavities. Sand–cement–water grouts are preferable for larger cavities. Measurements of grout quantities should be related to rock quality, and grouting pressures should be designed to force the grout into the fissures without exceeding the ground pre-stress (typically 1000 to 2000 kPa is used). With good control, hole spacings of

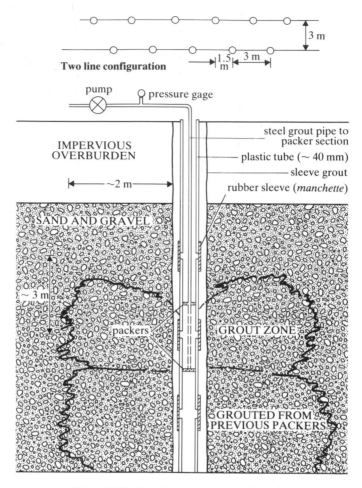

Figure 5.21 Overburden grouting schematic.

about 5 m should give complete coverage. Packer section spacings should be related to ground conditions but should not exceed 5 m. When grouting close to rockhead, closer hole spacings with reduced pressures may be necessary to prevent the grout from leaking out into the overburden contact. Effective grouting should reduce the permeability of jointed and fissured rock to about 10^{-10} m s^{-1}.

5.4 Dam settlements and distortion

Settlement and settlement rates are predicted from the results of consolidation tests by methods discussed in Chapter 2. Differential settlement analysis

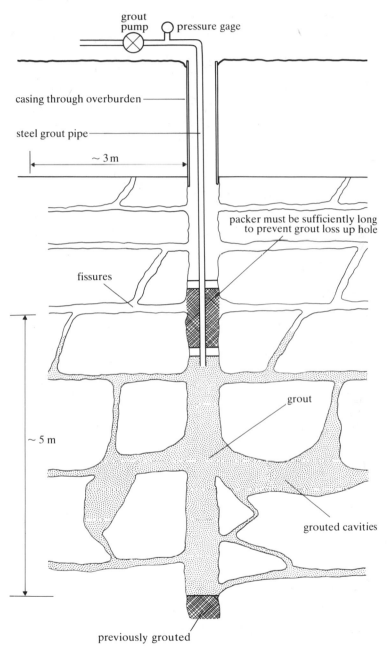

Figure 5.22 Rock grouting schematic.

is used to predict dam distortion in order to ensure that distortional strains are sufficiently low that cracking does not occur and that compressive stresses are, everywhere, substantially greater than the water pressure at that location. Consolidation and triaxial tests must be carried out on samples of all foundation soils and on compacted samples of the materials used in the earth structure. Where valley walls are narrow and steep, stress changes due to soil arching and the possibility of embankment rupture due to wall shear must also be investigated. Earth construction materials and associated features of earth dams must be engineered to be compatible with each other and with the physical properties and topography of the site. Settlements and distortions must be estimated in order that camber and curvature may be designed for, and movements that may affect spillway or drawoff structures may be anticipated.

5.4.1 Distortion due to foundation conditions

Figure 5.23 shows some common dam distortions due to foundation and abutment conditions. Foundation settlement (δ_F) on Figure 5.23a is calculated in the manner outlined on Figure 2.33. The embankment compression (δ_E) is given as

$$\delta_E = \int_0^H m_v \Delta\sigma_v = \int_0^H m_v z\gamma \, dz = m_v\gamma H^2/2$$

where m_v is the compression modulus of the compacted fill.

If it is now assumed that the settled dam base forms part of a circular arc, the average base strain due to elongation is given as

$$\epsilon_h = -\frac{\theta}{2\sin(\theta/2)} + 1$$

where θ is the angle, in radians, subtended by the arc length.

For moderate settlements, the base extensional strain can be estimated as

$$\epsilon_h = -\frac{2[(B/2)^2 + (\delta_F)^2]^{1/2}}{B} + 1 \tag{5.4}$$

The vertical compressive strain is given as

$$\epsilon_v = \frac{\delta_E}{H} = \frac{m_v\gamma H}{2} \tag{5.5}$$

Distortional strain under plane strain is given by $\epsilon_v - \epsilon_h$, and it should be determined that this distortional strain is less than that required to cause rupture in compacted samples tested under plane strain or triaxial conditions

(a) Base elongation (extension) and longitudinal cracking

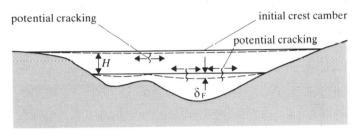

(b) Extension and transverse cracking in broad valley

(c) Arching and shear distortion in narrow valley

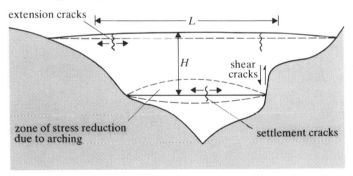

Figure 5.23 Dam distortion and cracking.

where the cell pressure is reduced as the axial stress is increased. It is also important to estimate the minimum minor principal stress in the interior base of the embankment, and this may be done by assuming a linear stress–strain relation in the form

$$\upsilon_h = K\gamma H - E_h \epsilon_h \tag{5.6}$$

where K is the coefficient of horizontal stress after compaction and may be measured in K_0 tests on compacted samples (usually about 0.5) and E_h can be obtained from triaxial extension testing on compacted samples. It should be determined that σ_h is significantly higher than the internal pore-water pressure. If the internal pore-water pressure exceeds the minimum principal stress at any point, cracking by hydraulic fracture will develop. Figure 5.24 shows the form of stress–strain behavior expected due to lateral movements in linear and work-hardening plastic materials, and it may be noted that plastic materials can withstand considerably more lateral compression than a linear material, but that their behavior in lateral extension is not so markedly different. Extensional strains should obviously be limited in earth dam design and construction.

Longitudinal differential settlements, as depicted on Figure 5.23b, can be analyzed in terms of distortion and stress changes following the methods outlined above. In this case the dam behaves essentially as a beam, and bending moments that induce tensile stresses either at the dam crest or dam base must be limited by forming abutment slopes and by downcutting any bedrock topographical highs (if exposed), or by foundation treatments to reduce the compressibility of adjacent soils. Transverse cracking is obviously more dangerous than longitudinal cracking because continuous channels in

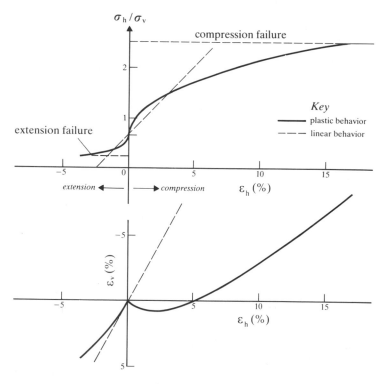

Figure 5.24 Linear and plastic deformations.

the direction of water flow may be formed. Such cracks can lead to piping erosion and breaching of the structure.

Figure 5.23c indicates a condition where arching can develop over a narrow span. If steeply dipping rock protrusions in the abutment areas are not shaped, high shear strains and possible rupture surfaces may develop. Tensile stresses close to the abutments can develop due to differential settlements within the embankment. Stress reductions (both vertical and lateral) can develop at the base of the dam with the combination of foundation settlement and embankment arching. Most of these settlement problems associated with high earth-fill dams constructed in narrow gorges can be avoided by correct abutment shaping, removal of all soft foundation materials and high compaction of the earth-fill materials. The potential for hdyraulic fracture is still most acute in this situation, however, and stress analysis must be carried out. As a first approximation, the Terzaghi (1943) arching equation

$$\sigma_v = \frac{L\gamma}{2K \tan \phi} \left\{ 1 - \exp[- 2K(\tan \phi)z/L] \right\} \tag{5.7}$$

can be used. In this equation L is the width of the valley (a variable in this case), K is a constant which may be approximated as $K = 1$ and z is the depth from the dam crest. When the dam height exceeds its chord length or average length $(H > L)$, the incremental vertical stress increases due to further increases in compacted dam height will diminish and, in the limit, the vertical stress will approach

$$\sigma_v = L\gamma/2 \tan \phi$$

which would be in the range of $L\gamma/2$ to $L\gamma$. Water pressures in the dam may approach the full reservoir hydrostatic value of $u = H\gamma_w$. To prevent hydraulic fracture the minimum stress must be greater than the water pressure. If the minimum stress σ_h is assumed to be $L\gamma/2$, this analysis would suggest that any dam with an average chord length less than the dam height would be a candidate for failure by hydraulic fracturing. However, there are a number of earth dams operating successfully with chord lengths less than the dam height, and Mica Dam, a 244 m high dam on sands and gravels overlying mica schists, has a height approximately equal to three times the chord length: the crest length of Mica Dam is 780 m. A concrete gravity arch dam would have been a viable alternative for this site.

5.4.2 Internal distortions in zoned dams

Differential settlements between the zones in a large earth dam can substantially alter the stress distribution within the dam, particularly when a

195

vertical soft core is used with a stiff shell material. The type of settlement profile that might be expected in this case is shown on Figure 5.25a. A linear idealization in which the distortion is assumed to occur only in the filter zone is shown on Figure 5.25b and can be analyzed as follows.

Differential movement between the core and shell at depth z is estimated as

$$\delta = dz\, \frac{\sigma_{vz}}{D_c}$$

where D_c is the confined modulus of the core. Shear distortion in the filter zone at depth z is given as

$$\epsilon_z = \delta / t$$

Shear stress developed in the filter zone at depth z is then

$$\tau_z = G\epsilon_z = \frac{G\delta}{t}$$

The resulting vertical core stress at depth z is calculated as

$$\sigma_{vz} = z\gamma_c - \frac{2\tau_z dz}{w}$$

Then

$$\sigma_{vz} = \frac{z\gamma_c}{1 + \dfrac{G}{D_c}\dfrac{z^2}{tw}} \tag{5.8}$$

where G, t and w are shown on Figure 5.25b.

The shear modulus of the filter material can be found from simple shear testing or estimated from pseudo-elastic values of (E, ν) obtained from drained triaxial testing. The constrained modulus of the core material is obtained from consolidation testing as the inverse of the volume compression index (m_v). At the dam base (where $z = H$) the vertical core stress would be reduced substantially below the gravitational value of $H\gamma_c$ when the core and filter widths are small compared to the dam height, or when the shear modulus of the filter is the same order of magnitude as the constrained modulus of the core material. In order to ensure that the minimum total stress in the core is higher than the core water pressure during dam operation, the denominator in Equation 5.8 should be less than $(K_0 H\gamma_c)/(H - F_b)\gamma_w$ and for normal freeboards the term $Gz^2/D_c tw$ would generally be less than 0.35. Greater core compaction, wider cores and thicker, less compact filter

196

(a) Differential settlement between shells and core

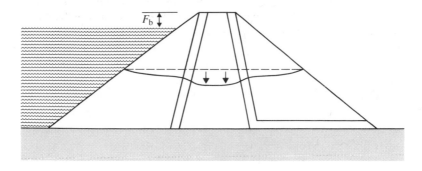

(b) Idealized distortion

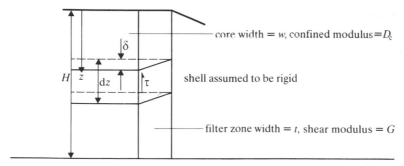

Figure 5.25 Core–filter–shell distortion.

zones will assist in reducing the potential for hydraulic fracturing due to core stress reduction upon settlement. The problem is accentuated when impervious cores are placed in a wet state and do not consolidate during the construction period. Transverse core distortion (simple shear) also develops when the downstream shell materials distort to resist the water pressures transmitted by the core. To limit this distortion to less than 1% of the dam height the shear modulus (G) of the shell material should exceed the quantity $200H\gamma_w/\cot \beta$. Since the distortion will be greater in the central portion of a long dam than near the abutments, such dams are often constructed with an upstream arch so that the upper part will be in compression after the reservoir is filled.

In order to improve the estimation of stresses and distortions within earth dams, particularly within zoned dams where the material properties of various zones are markedly different, finite element methods (FEM) have been applied (Eisenstein *et al.* 1972, Milbury & Duncan 1972, Seed *et al.* 1975a). Using FEM it is possible to incorporate the time deformation relations found from laboratory testing into the analysis and to predict the equilibrium stress distribution and deformations with some confidence.

Non-linear elastic and elasto-plastic models have been developed for application to zoned dams (Naylor 1975) and these models also assist in evaluating construction safety factors by predicting the internal pore-water pressures likely to develop during construction. Differential settlements between cores and shells of quite different properties, which is believed to have led to hydraulic fracturing in two large earth dams, can be analyzed more rigorously using FEM (Seed *et al.* 1975a). A three-dimensional FEM analysis of Mica Dam (Eisenstein & Simmons 1975) indicated that the usual two-dimensional analysis cannot accurately evaluate the potential for cracking near the abutments.

In an attempt to provide some guidance to field performance, Justo (1973) reviewed numerous case records and produced a table indicating dangerous settlement levels for various dam considerations. Table 5.3 is a modification of this type of practical guide.

5.5 Earthquake and rapid drawdown design

The equivalent static slope stability analysis outlined in Section 4.4.2 can be applied to obtain a safety factor for earth dams under earthquake loadings. This static approach, however, gives no indication of possible distortions or stress changes during earthquake loadings and numerical dynamic analysis is recommended for any large earth dams to be located in an area where there is a probability of earthquake occurrence. Where only static analysis is carried out, the values of seismic coefficient should be selected with reference to earth dam experience and embankment material properties (Seed & Martin 1966, Seed 1967).

The dynamic response approach (Newmark 1965) assumes a rigid plastic embankment material and determines displacements that would develop if a yield acceleration were exceeded on a failure surface. Seed (1966) introduced an alternative approach using stress analysis to determine the initial embankment stresses and the stress changes due to earthquake loading on the embankment. Deformation predictions are obtained by subjecting laboratory samples to the stress sequences found from the analysis. Both of

Table 5.3 Dam settlement and cracking.

Max. post-construction settlement/$H \cot \beta$	Cracking problems
0.001	no cracking
0.002	thin reinforced concrete facings may crack
0.003	longitudinal cracks in dry cores
0.005	transverse cracking of cores and oblique core-to-shell cracks
0.01	longitudinal cracks in wet cores; jointed concrete facings crack
0.02	danger of transverse cracking and piping failures

these approaches have practical and theoretical drawbacks. Back-analysis of failures of earth structures under earthquake loadings (Seed *et al.* 1969, 1975b) and large-scale field tests (Hemborg & Keightley 1964) have provided further insight into the behavior of earth masses during earthquakes. Recent dynamic finite element models have been developed for predicting dynamic behavior of earth dams (see, for example, Mizukoshi & Mimura 1975, Seed *et al.* 1975a, Watanabe 1975). Mizukoshi and Mimura indicate that horizontal accelerations will be 2 to 4 times higher at the dam crest than at the foundation, and higher at exposed surfaces than in the central core. They indicate that a variable seismic coefficient for static analyses will reduce the factor of safety by about 0.2 from the constant coefficient analysis. An increased crest width is recommended for dams in earthquake-prone areas so that the internal portions of the embankment will be better protected from dynamic accelerations.

Watanabe (1975) used model studies and field observations to establish correlation with a numerical seismic analysis for rock and earth dams. Centrifugal models with superimposed cyclic accelerations are also being developed and can be used to study dams under earthquake loadings (Schofield 1981).

The rapid drawdown design method outlined in Section 4.4.3 can be applied to a homogeneous earth dam and to the failure mechanisms shown on Figure 5.7. The mechanisms on Figure 5.7 may be critical under drawdown conditions and are often analyzed using undrained strengths (short-term). Barrett and Moore (1975) suggest that an undrained effective stress path approach should be used to evaluate the response of zoned dams under drawdown. They indicate that $\bar{B} = 1$ (equivalent to $r_u = \gamma_w/\gamma$) may be non-conservative in clay cores since additional pore-water pressures may be caused by distortion. From piezometer measurements on Tooma Dam, Pinkerton and McConnell (1964) calculated a value of $\bar{B} = 1.5$ in the steep core during a 5.5 day drawdown. Other case records show $\bar{B}$ varying from unity for a steep saturated core (Glover *et al.* 1948) to less than 0.5 for a sloping 80% saturated core (Bazett 1961). When rapid drawdown is a major design feature of an earth dam it is recommended that effective stress path analysis of pore-water pressure generation (Morgenstern 1963, Barrett & Moore 1975) be carried out or, alternatively, that centrifugal model studies be conducted (Avgherinos & Schofield 1969).

5.6 Some special considerations in construction of earth dams

Although excavation of overburden soils is recommended whenever it is economically feasible, some weaker rocks (particularly shales) may become softened under the vertical stress reduction due to overburden removal. Heave cracking and possible lateral movement on weak layers within the

199

rock structure can occur when the lateral stresses in the rock are sufficiently high. In such cases it may be difficult to prepare a suitable foundation for the dam. The character of weak foundation rocks should be investigated prior to excavation by conducting *in-situ* stress measurements, by examining cores for weaknesses, and by testing rock cores for expansion and strength after wetting and drying cycles.

It is preferable to carry out dam construction in the dry, although cutoffs and the lower portions of some large dams have been successfully completed by underwater placement and compaction of materials (see, for example, Golder & Bazett 1967, Guilford & Chan 1969). For smaller dam heights where spillway structures can be founded on competent rock in the river bed, temporary river diversion around the spillway construction area can be accomplished using cofferdams, and water is then diverted through the spillway during dam construction. Cofferdams are also used with or without cutoffs as temporary dams to maintain river levels on each side of an abutment spillway. In the latter case cofferdams constructed with rockfill are usually incorporated in the dam as shown on Figure 5.26. Sutcliffe (1965) describes the construction of a till cofferdam in 5 m of water with a flow rate of about 1 m s^{-1}.

Earth dam materials are usually compacted in lifts ranging between about 0.3 m and about 0.6 m depending on the type of material and compaction equipment. The maximum particle size should be less than about 0.6 times the lift thickness. The moisture content of shell materials should be close to optimum to promote uniform compaction and reduce segregation, but water contents less than optimum are used if the material is sufficiently fine-grained that excess pore-water pressures would be created and maintained during construction. Goel and Das (1981) present a case study on construction pore-water pressures, and the general prediction of pore-water pressures in earth dams is discussed in detail by Richards and Chan (1969). Compacted dry densities may be close to or exceed the standard Proctor maximum where foundation settlements will be relatively small, but the degree of compaction may be controlled to provide a more flexible embankment where relatively high foundation settlements are anticipated. On large earth dam projects where high-capacity bottom dump trucks or conveyors are used for material placement the rate of placement may easily exceed 10^5 m^3 per day. Strict supervision together with continued monitoring of in-place densities is necessary to ensure uniformity of compaction. The moisture content of core materials should be carefully controlled to ensure a uniformity of permeability, strength and flexibility within the compacted core. Water contents 1 to 2% wet of optimum are often used in order to control the uniformity of the resulting density better. A wet fill method is sometimes used (water contents 4 to 6% above optimum) with silty core materials to produce better flexibility or to facilitate placement in wet conditions (Bernell 1964, Nilsson & Lofquist 1955). Construction pore-water pressures in the

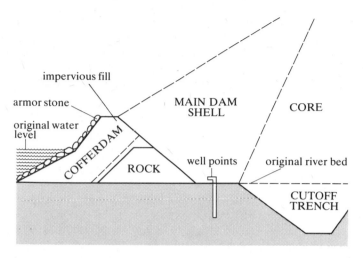

Figure 5.26 Dam construction with cofferdams.

core will be higher with wetter compaction, and some dam cores are compacted 1 to 2% dry of optimum to reduce construction pore-water pressures (Collins 1964). Wilson and Squier (1969) suggest that the lower portion of the core be compacted dry of optimum to reduce pore-water pressures and compressibility, but the upper portion and zones adjacent to abutments be compacted wet of optimum to retain greater flexibility. Compacted dry densities of core materials range from about 90 to 98% of standard Proctor depending on the material and the expected distortions. Table 5.4 lists some considerations and typical compaction control on core materials. Filter materials must be compacted to densities such that no shrinkage will occur upon further wetting, and should be compacted to greater densities to resist lateral compression under the water pressure gradient exerted by the core. Relative densities in excess of 70% are considered satisfactory for transition zones (Wilson & Squier 1969).

Rockfill can be suitably compacted in lifts up to 2 or 3 m (Cooke 1960) and is normally compacted using vibratory rollers of up to 15 tonne static mass. Some details of field density test procedures on coarse fill materials are summarized by Wilson and Squier (1969).

5.7 Monitoring, performance and maintenance of earth dams

The factor of safety of an earth dam may be lowest during construction, during first reservoir filling, during certain operational conditions or after decades of service. It is essential that instrumentation is incorporated in a large dam to monitor its performance from the initiation of construction through its maintained service life. Typical instrumentation types and loca-

Table 5.4 Compaction of core materials.

Material type	Resistance to piping	Resistance to cracking	Typical compaction	Importance of moisture control
GM SM < 6% clay	low but increases with compaction and plasticity	low but increases with lower compaction and plasticity	40 to 80 tonne; pneumatic rollers; 95% standard Proctor	good control required to maintain flexibility
GC SC < 20% clay	intermediate resistance	intermediate resistance at typical compaction	40 to 80 tonne; pneumatic rollers; 98% standard Proctor	only to control construction pore-water pressure
CL	high resistance	sufficiently flexible to resist higher settlements	pneumatic or sheepsfoot rollers; 92 to 98% standard Proctor	moderate control required to control pore-water pressures
CH	high resistance with good compaction	very flexible to resist large settlements	sheepsfoot roller; 92 to 98% standard Proctor	may be compacted dry of optimum to avoid high pore-water pressures

tions are indicated on Figure 5.27 and the purpose of various instruments is outlined below.

(a) Piezometers, installed in the foundation, embankment and core, are used to monitor the build-up and dissipation of pore-water pressures during construction, so that stability of the embankment and the progress of the foundation consolidation can be calculated at any time. The most critical section (usually the central section) and, in a long dam, several other sections should be instrumented. Piezometers should be selected for their durability and ability to monitor the required pressures reliably (Bishop *et al.* 1961).

(b) Settlement gages are installed in the foundation and dam to monitor the settlements during construction and reservoir filling. These should be located in a number of selected sections along the dam length so that longitudinal distortion can also be evaluated. Hydraulic settlement gages (Bozozuk 1969) can be installed along the dam axis to measure distortion accurately while spiral foot settlement rods are used to monitor vertical settlements and differential settlements. Figure 5.8 shows a hydraulic settlement gage installation in a cutoff trench.

(c) Inclinometers are installed in the downstream shell and in the foundation, to measure the lateral deformation due to foundation yielding and the shear strain developed in the shell in resisting the water pressure of the reservoir. Lateral downstream displacements of about 1% of the dam height may develop when the reservoir level is raised. Magnetic settlement rings can be installed around the inclinometer tubes to measure differential vertical settlements.

(d) Piezometers are placed in the core and downstream shell to monitor the operational pore-water pressures and the effectiveness of filters. These data are necessary for stability analysis. Piezometers in the upstream shell and foundation, if of sufficiently small compliance, will monitor the pore-water pressure changes during drawdown, while piezometers in the downstream foundation monitor the effectiveness of the cutoff and provide data to calculate uplift forces and seepage gradients.

(e) Accelerometers should be installed near the crest of the dam, and in the downstream slope when the dam is constructed in an earthquake-prone area.

(f) Total stress cells should be installed in all areas of the dam where distortion may cause a reduction in compressive stresses. Hydraulic or pneumatic stress gages are generally more durable than strain-gaged cells, although vibrating-wire cells have been found to perform satisfactorily.

(g) Extensometers may be used to monitor lateral movements within or between zones of a zoned dam, but are usually incorporated only in conjunction with numerous pressure cells to study the stress strain properties of the compacted earth. Wilson and Squier (1969) describe the deformations measured in several earth and rockfill dams using detailed instrumentation.

Alarm systems can be connected to piezometers and drains to warn operators of potential water pressure or seepage problems. An instrument

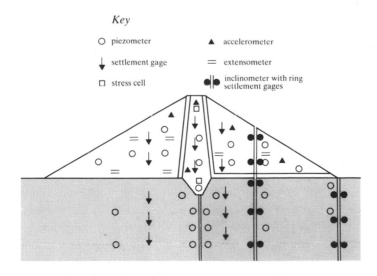

Figure 5.27   Typical instrumentation in a large dam.

house is often provided for monitoring of all hydraulic, pneumatic and electrical instrumentation. Hand compaction of materials around the instruments is necessary to maintain uniformity and prevent damage to the instruments from equipment. In heavy construction operations, some ingenuity in installation and protection of instruments is usually necessary to obtain a reasonable survival rate. Equipment operators should be advised of the importance of these instruments in the monitoring and control of dam operations over its lifetime. Instrumentation should also be included (piezometers and slope movement indicators) in reservoir slopes if there is any reasonable doubt about the stability of these slopes. The instrumentation of the 220 m high Oroville Dam (USA) is described by O'Rourke (1974), and the stresses and movements in this dam are detailed by Kulhawy and Duncan (1972).

All dams, large and small, should be subjected to periodic engineering inspections (usually twice a year or more frequently) for any evidence of deviations in crest alignment, cracking or bulging of slopes, sloughing or erosion, animal burrows and seepage (springs or boils). Instrumentation installations should be monitored on a regular basis between inspections by dam operators as well as by the inspecting engineers. Inspection activities in various countries have been summarized by ASCE (1974b). Any increase in water pressure in the dam or evidence of seepage should be recognized as a distress signal requiring immediate remedial action: immediate reservoir lowering may prevent a disaster and this should be followed by placement of a toe filter berm, grouting to seal cracks or installation of relief wells as permanent remedial measures before refilling. Dam failures have been related to lack of an adequate emergency spillway and to:

(a) extremely poor site conditions, particularly with regard to adverse seepage problems and poor foundation conditions;
(b) inadequate foundation preparation, particularly with regard to water-tightness;
(c) inadequate construction control resulting in high construction pore-water pressures and failures in embankment slopes;
(d) inadequate monitoring, inspection and maintenance resulting in piping failures;
(e) earthquakes or fault movements – impoundment reservoirs can create subsurface water pressures sufficient to trigger movement in previously stable fault zones.

Although every earth dam project begins with an engineering design, prominent earth dam engineers stress that design is a continuing process with laboratory materials testing, stability analysis and deformation analysis continuing as field data become available during dam construction (Peck

1969b). This can lead to design improvements and cost savings on large earth dam projects.

Many older dams suffered failures due to inadequate design or construction, and adverse foundation seepage left a few unable to impound water to the design levels (Walters 1971, ICOLD 1973, Thomas 1976). Much has been learned about dam buildings during the past half-century but dam failures have occurred during that time, and the case records briefly outlined on Tables 5.5 and 5.6 can provide the reader with detailed insight into the variety of problems encountered in earth dam design and construction.

5.8 Mine-tailings dams and process-water impoundments

In response to the need for safer, more efficient and environmentally acceptable methods of designing, constructing and monitoring impoundment structures that retain mine wastes or process waters, geotechnical engineers have become increasingly more involved in these aspects of resource development during the past two decades. Earth structures engineering expertise can contribute substantially to the design of tailings dams, process-water impoundments, solid waste dumps and mine backfill. The latter two areas of interest are discussed in Chapters 4 and 6, respectively.

5.8.1 Mine-tailings dams

In some semi-arid regions the finer grain sizes from mill tailings (sizes less than 0.02 mm and commonly referred to as slimes) can be used for dam building because they have significantly dry strength when sunbaked. In more humid climates the coarser grain sizes (having D_{10} generally greater than 0.05 mm and referred to as classified tailings sands) are used for dam building, and the slimes are pumped out into the tailings area creating a pond remote from the dam. This is done mainly because the high moisture retention of the slimes makes compaction difficult and the resulting dam would be weak and susceptible to liquefaction. While reasonably free-draining, the classified sands are also difficult to compact because they are poorly graded, having uniformity coefficients generally less than 6. Mine tailings are produced over the life of the mine and, as a result, the tailings dams often begin as small starter dams and grow to great heights during the operational life of the mine. Abandoned tailings dams must, however, continue to impound the fine tailings and surface waters that may collect in the pond, and if the dam is not correctly designed, protected and maintained it is subject to failure in the long term, particularly under flood conditions, earthquake loadings or as a result of long-term erosion. An early method of construction, which was suited to damming natural basins using only tailings

Table 5.5 Some case studies of dam and reservoir failures.

Name of project	Problem and resulting failure	Reference
South Fork Dam (USA): completed 1853, failed 1889; 22 m high	Inadequate spillway resulted in overtopping. Reservoir emptied in less than one hour with loss of 2200 lives	Thomas (1976)
Eildon Dam (Australia): failed 1929; 43 m high	Failure of shell to support clay core under drawdown resulted in slumping of upstream shell. Rockfill reconstructed to concrete core wall	Knight (1938)
Oros Dam (Brazil): failed 1960 during construction; 54 m high	Construction delay resulted in overtopping and erosion of nearly 1×10^6 m^3 of earth. Evacuation just prior to failure	ENR (1960)
Vajont Reservoir (Italy): slope failure 1963; 265 m high	Creep movements on weak planes in steeply dipping Cretaceous limestones began upon reservoir filling but suddenly accelerated to dump 250×10^6 m^3 of rock into reservoir overtopping dam and destroying several villages	Kiersch (1964), Kenney (1965), Muller (1964b)
Morwell Reservoir (Australia): failed 1966 after fast filling of reservoir	Piping failure resulted from the compaction of dispersive clay at less than optimum water content. Two holes of about 0.3 m diameter formed by piping through the homogeneous embankment	James and Wickham (1970)
Baldwin Hills Reservoir (USA): failed 1963; constructed 1951, 80 m high	Constructed in an area previously subjected to subsidence and faulting from oil extraction and tectonic forces. The asphaltic lining cracked and the dam was breached by piping. Helicopter rescue and evacuation limited loss of life, damage exceeded $12 million	Jansen *et al.* (1967), Leps (1972)
Cascade Dam (Australia): completed 1926, failed 1929; 19 m high	Rockfill dam of 3 to 5 tonne granite behind a concrete face dislodged due to impact of floodwater waves	Thomas (1976)
Sheffield Dam (USA): constructed 1917, failed 1929; 8 m high	Earth dam of silty sand with upstream blanket failed during earthquake due to liquefaction of sand embankment	Seed *et al.* (1969)
San Fernando Dam (USA): constructed 1915, failed 1971; 44 m high	Liquefaction failure under unexpected earthquake of Richter magnitude 6.6. Lower part of dam constructed by hydraulic filling	Thomas (1976)
Buffalo Creek (USA): filling behind rock dump (1960) to 45 m height, failed 1972	A non-engineered tailings dam failed due to improper foundation and zoning. Failure involved slumping and liquefaction of loose wastes. 125 lives lost	Davies (1973)
Teton Dam (USA): failed 1976 after reservoir filling; 100 m high	Differential settlement and possible hydraulic fracturing of silty core cutoff material together with fractures in abutment rock caused piping failure	USDI (1977)

materials, is called the upstream method and is shown on Figure 5.28a. As the dam height is increased, the critical circle will extend into the weaker uncompacted materials and the risk of failure increases. This type of construction is not suitable in earthquake-prone areas because the saturated silts are likely to liquefy. Indeed, this method of construction is not generally recommended for heights in excess of 8 m. Downstream or centerline methods as shown on Figures 5.28b and c produce much safer earth structures, but additional fill material is necessary. When waste rock is used for the additional fill, as shown on Figure 5.28b, a filter zone of crushed rock and gravel must be designed and placed to prevent piping of the sands into the course rockfill. When relatively impervious earth borrow is used, as shown on Figure 5.28c, it should be compacted as a core, with the classified sand being used for the downstream shell.

Leakage through a tailings dam can be a major consideration in the operation of the tailings pond, since it is more efficient and environmentally more acceptable to recirculate mill process water. When the leakage is a small percentage of the circulating water volume, the upstream pond can be maintained and clear water can be returned through decant culverts or filter-wrapped decant siphon pipes. These types of overflows are less susceptible to damage from tailings flow or ice action than concrete decant towers. Pumps on floating barges are sometimes used. If the leakage is large, recirculating water may be retained by creating a pond on the downstream side of the dam. In remote areas these reclaim ponds are often small low-lying natural lakes that have been approved for use as mill water sources (see, for example, Fig. 1.25). There are innumerable possible variations in design but there are two major practical considerations that tend to keep the designs fairly simple: large capital investment in dam construction is not generally justified over the relatively short lifespan of operations, and material segregation in the pond can make seepage quantities very difficult to predict. Tailings slimes, with permeabilities below 10^{-7} m s^{-1}, can create an upstream blanket over parts of the pond and effectively prevent seepage. Some novel designs have been proposed for tailings impoundments on terrain where suitable natural basins are not available for disposal (see, for example, Shields 1975, Robinsky 1975, 1978).

A large number of factors can affect the design and operation of tailings disposal areas, the major considerations being the following.

(a) Type of waste material and proximity to developed areas: these two factors are interrelated because pollution of ground or surface waters is of less public concern in remote areas. In developed areas it is often necessary to seal tailings ponds to prevent groundwater contamination, and to treat waste waters before effluents can be discharged on the surface. Surface stabilization of disposal areas is usually mandatory in most areas.

207

Table 5.6 Some dams requiring special construction or remedial measures.

Name of project	Problem and special construction or remedial measures	Reference
Victor Braunig Dam (USA): completed 1963, embankment failure 1969; 20 m high	Clay embankment cracked and settled due to shear failure in highly plastic clay underlying toe of embankment. Stabilizing berm constructed and dam repaired	Reuss and Schattenberg (1972)
Gepatsch Dam (Austria): completed 1966; 153 m high	During initial filling, movements were observed in rock slopes above the reservoir due to increased water pressure. Movements were controlled by controlling the reservoir level	Lauffer *et al.* (1967)
El Bosque Dam (Mexico): completed 1954; 70 m high	High leakage necessitated reservoir lowering. Sink holes discovered at upstream toe indicated piping in foundation. Grout curtain installed to reduce seepage	Mooser (1964), Marsal and Resendiz (1971)
Duncan Dam (Canada): completed 1967; 36 m high	Transverse cracking developed during construction due to high settlements (up to 2.5 m close to abutment) in compressible foundation materials. Cracking controlled by good construction control and slope flattening	Gordon and Duguid (1970), Eisenstein *et al.* (1972)
Hills Creek Dam (USA): completed 1962, leakage increased in 1969; 104 m high	High leakage developed through core of compacted alluvial material containing plastic fines. Core grouting to seal voids in the core and prevent piping of core fines was used to correct the problem	Jenkins and Bankofier (1972)
Laguna Dam (Mexico): constructed 1908, failed 1970	Catastrophic piping failure in abutment of weathered volcanic tuffs and basalts. Dam repairs and slurry trench cutoff installed to control seepage through residual foundation materials	Marsal and Pohlenz (1972)
Mica Dam (Canada): completed 1973; 244 m high	Rigid specifications for foundation and abutment preparation and for material selection and placement were necessitated to avoid cracking (due to the steepness of the valley walls) and to avoid potential earthquake damage	Webster (1970)
Lower Svir Dam (Russia): failed during construction 1935; 28 m high	Liquefaction failure of uniform material (70% between 0.3 and 1.2 mm) at 4 : 1 to 2.5 : 1 slopes initiated by nearby blasting during construction	Thomas (1976)
Mississippi Levees (USA): 1910 to 1940; variable height to 7 m	Landslide berms, riverside blankets and relief wells designed to control underseepage in alluvial deposits underlain by pervious sands	Turnbull and Mansur (1961)

(b) Seismic risk: having relatively high uniformity coefficients and being hydraulically transported, tailings disposal areas are subject to liquefaction under earthquake accelerations. Earthquake-resistant designs must be employed in seismic risk areas.

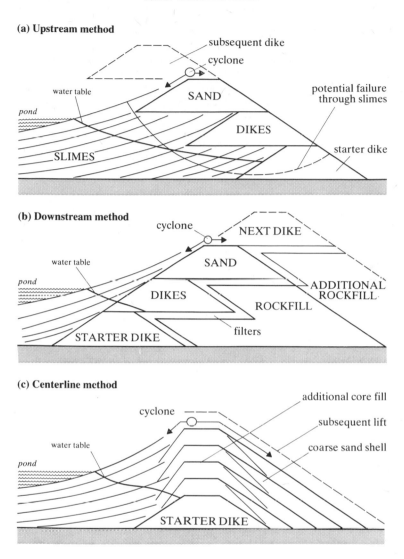

(a) Upstream method

subsequent dike

cyclone

water table

SAND

potential failure
through slimes

pond

DIKES

starter dike

SLIMES

(b) Downstream method

cyclone

NEXT DIKE

water table

SAND

pond

ADDITIONAL
ROCKFILL

DIKES

ROCKFILL

STARTER DIKE

filters

(c) Centerline method

additional core fill

cyclone

subsequent lift

water table

coarse sand shell

pond

STARTER DIKE

Figure 5.28 Tailings dam construction.

(c) Type of mine operation: the type and volume of mine wastes to dispose of on the surface depends on the type of mine operation. The extraction from potash mines, for example, is sufficiently high that the waste materials could all be disposed of in mine openings. In metalliferous mines, uranium mines and oil sands, the volume of wastes is too large for the mine voids and additional surface disposal is necessary. In addition, the grain sizes suitable for backfill are generally limited because of the drainage requirements during hydraulic filling operations. Open

Table 5.7 Mine waste disposal considerations.

Mine operation	Wastes and disposal method	Major geotechnical concerns (excl. costs)
Totally underground: due to proximity to populated area	Mill rejects are the only surface waste. As much as possible is put underground as backfill. Remaining slimes are disposed of on surface. Classified sands stored on surface are repulped for final void filling	Ground subsidence. Environmental contamination. Total reclamation of slimes pond
Totally underground: due to depth of ore body or surface water but in remote area	Mill rejects are the only surface waste. May be used as backfill or disposed of on surface. Average about 50:50, but could be up to about 80% either way under unusual circumstances	Backfill drainage. Suitable stable surface disposal. Environmental if toxic waste
Open pit (often with later underground option): remote from populated areas	Overburden from mine stripping often used for dike construction. Waste rock used for roads, dams and sometimes as backfill, remainder as waste rock dump. Mill rejects disposed of on surface and may later be used for backfill	Control surface runoff. Stable site for waste rock dump. Tailings dam and large surface tailings pond. Environmental if toxic wastes. Pit slopes

pit and underground operations have different requirements, as briefly summarized on Table 5.7.

(d) Overall mining costs: many metalliferous mines are high tonnage operations with a small profit margin per tonne during early operations due to capital amortization. Waste disposal is a significant factor in financial planning, and the economic use of mine wastes as backfill or for constructing roads, dikes or dams can increase the mine profit margin.

All of the factors discussed in this chapter should be considered in the design, construction and maintenance of tailings dams. In addition, the effects of any leaching of mill water on the embankment material or on the surrounding watershed must be considered. The thought that mine wastes are not suitable for dam construction is belied by the fact that the 91 m high Llyn Brianne water storage dam in Wales was constructed utilizing large quantities of mine wastes. Coarse tailings materials produced by spigotting, sluicing or by cyclone separation must be allowed to drain to a moisture content close to optimum and then must be spread and compacted to form the dam. Unsaturated sands that are placed without correct compaction can form steep slopes that have a high risk of failure, as shown on Figure 5.29.

The only major concession in earth dam principles as applied to tailings dams is in regard to the foundation design and treatment. Because the pond will have a slimes blanket and silty sand infill to reduce seepage pressures,

210

the risk of piping beneath a tailings dam is considered very low. Where such ponds are located on soft alluvial materials which are not capable of supporting a high dam but which cannot be economically excavated and replaced, controlled displacement construction of a starter dike may be approved (see Section 3.15). Figure 5.30 shows a cross section of a waste rock displacement dike. Careful construction control (knowing the depths of soft soil, knowing the rock tonnages dumped, calculating the displaced volumes, toe blasting and forward excavation as necessary to assist displacement, and not advancing over frozen soils) will ensure a stable dike from which a sand dam can be constructed.

Equivalent static analysis has often been used to analyze the resistance of tailings dams to earthquake loadings (CANMET 1972) but this method does not consider liquefaction potential. Marcuson *et al.* (1979) report an earthquake-induced failure of a tailings impoundment structure which had been designed using equivalent static analysis. Dobry and Alvarez (1969) provide additional details on the conditions causing seismic failure of several tailings dams. Schnabel *et al.* (1972) and Klohn *et al.* (1978) outline dynamic methods of analyzing tailings impoundments, using ground motion characteristics and material response analysis based on cyclic laboratory test data. Numerous researchers have suggested that a threshold level of ground shaking must be exceeded to cause significant damage to tailings dams, and a summary of this work indicates threshold levels between 0.1g and 0.3g (Conlin 1982). Threshold level would appear to depend on the duration of

Figure 5.29   Failure in oversteepened sand slope. (Photo courtesy of John D. Smith Engineering Associates Ltd.)

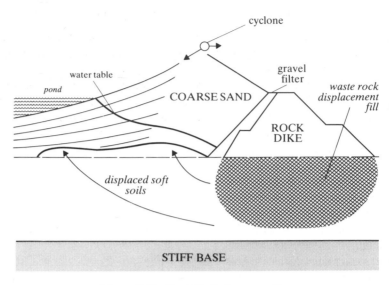

Figure 5.30 Rockfill displacement dike.

ground shaking and on the depth and *in-situ* properties of the foundation materials.

5.8.2 Process-water impoundments

Process waters used in the mining industry are not generally toxic but often contain sufficiently high levels of contaminants (sulphur, salts, radioactive isotopes, etc.) that they must be contained to avoid the contamination of ground water or treated before discharge to watersheds. Containment of process waters usually requires some form of blanket or liner over the entire pond area. Concrete and asphaltic liners have been used for decades in relatively small freshwater supply reservoirs to prevent leakage, but these types of liners are expensive and are not generally resistant to chemical deterioration. Compacted clay liners and polymeric membranes would appear to be generally more suitable for process-water impoundments.

The compacted permeability of a clay liner (usually 10^{-8} to 10^{-10} m s^{-1}) and the possible effects of the process water on the clay material (swelling or shrinking due to ion exchange or leaching) should be investigated in the laboratory. Compacting slightly wet of optimum using sheepsfoot rollers to produce a kneading action has been found to produce a more uniform, less fissured and lower permeability liner (Seed & Chan 1959, Mitchell *et al.* 1965). In the absence of a suitable natural clay (25% or more clay sizes with no large stones to interfere with compaction) bentonite and sand mixes will produce a liner of similar characteristics. Uniformity of material and uniform compaction are important field control considerations. Kisch (1959) presents

212

detailed theoretical considerations with regard to seepage from clay-lined ponds, and the effects of chemico-osmosis, organic leachates and attenuation of pollutants are, respectively, considered by Mitchell *et al.* (1973), Anderson and Brown (1981) and Griffin and Shimp (1978).

Leakage from polymeric liners such as butyl rubber, neoprene or polyvinyl chlorides is theoretically limited to that due to diffusion, osmosis and liner solubility but, in practice, the leakage through holes and poor seams is likely to be more important. Good construction control before and during installation of the membrane and prevention of membrane damage during operation are important considerations. From field measurements published by Kays (1977), Folkes (1982) has estimated that the apparent permeability of correctly installed membrane liners ranges from 5×10^{-12} to 5×10^{-14} m s^{-1} for heads of about 6 m. Before selecting a particular type of membrane the engineer should investigate the chemical resistance of the material to the containment fluid (Haxo 1981).

Where the natural groundwater table is close to the surface, the seepage or leakage from the pond must be reduced to a level where dilution with the ground water will not produce any unacceptable change in groundwater quality. Groundwater flow and groundwater quality prior to impoundment should be monitored, and sampling wells should be installed on the downstream side of the pond, as shown on Figure 5.31a, to monitor operational water quality. Where the groundwater table is remote from the ground surface and the subsoils are fine-grained, partially saturated or relatively impervious, there may be a significant time delay before process-water leakage is detected in observation wells. This time delay may be of great importance with regard to certain contaminants, but the processes of contaminant movement through partially saturated soils are complex and the delay time is difficult to assess accurately. Folkes (1982) presents detailed discussion of the processes and considerations required for the design and monitoring of liners under this condition.

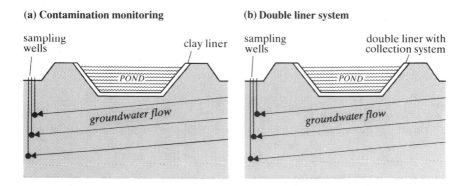

(a) Contamination monitoring

sampling wells   clay liner

POND

groundwater flow

(b) Double liner system

sampling wells   double liner with collection system

POND

groundwater flow

Figure 5.31 Process-water impoundment.

Where process water is particularly toxic it may be necessary to devise a double liner system with a filter zone between the two liners, as shown on Figure 5.31b. Leakage to a filter zone between two liners can be collected by gravity drainage to a sump and pumping back to the pond, thus reducing the gradient across the lower liner to a negligible value. Alternatively a hydrostatic pressure slightly lower than the minimum pond level could be maintained in the filter, and the filter water could be exchanged with fresh water on a regular schedule, thus reducing the concentration of contaminant in the water seeping through the lower liner. Problems associated with water exchange operation may, however, negate any benefits of this alternative. A third alternative in the operation of the double liner system would be to maintain fresh water in the filter at a small excess pressure, thus creating a freshwater gradient into the pond as well as into the subgrade. This type of operation would require that a sand or gravel filter be placed over the upper liner to prevent piping or possible blowout of a soil liner or floating of a membrane liner. Freezing of water must also be considered in colder climates. For an in-depth discussion on the subject of control of contaminant migration by the use of liners, the reader is referred to Folkes (1982).

5.9   Problems on earth dams

Problem 1   Part of a proposed hydroelectric power reservoir will extend over an old buried channel flanked on both sides by competent impervious glacial till. A dike must be constructed at the location shown on Figure 5.32 to impound the reservoir on the downstream end of the old channel. The old channel deposits (sands and gravels) taper out about 2000 m upstream of the dike and about 1000 m downstream of the dike location due to changes in bedrock elevation (6 m average thickness). Figure 5.32 also shows test data on the only suitable earth construction material available locally for dike construction. A limited supply of sand and quarry rock is available for filters and rip-rap.

(a) Using the information provided, design a stable impoundment dike founded on the till strata at the proposed location. Use a static design with $F = 1.3$, assuming that the dike must be stable ($F > 1$) under the condition of rapid drawdown. Design the downstream filter and specify appropriate grading characteristics for the filter sand.
(b) Check the approximate operational safety factors for your design if the dike is constructed in earthquake zone 2, and note any design changes that you would recommend if this earthquake condition was to be anticipated.
(c) Calculate the leakage from the reservoir in the area of the dike and the underseepage pressures. Comment on the suitability of the proposed

(a) Plan and elevation along buried alluvial channel

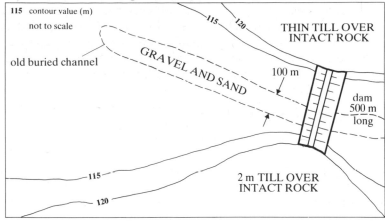

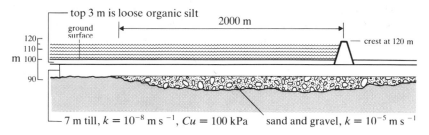

(b) Test results of samples of borrow material for dam

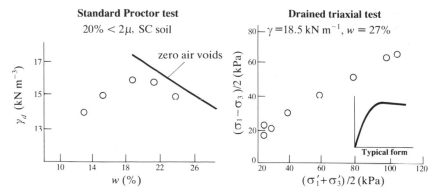

Figure 5.32 Reservoir leakage and uplift.

design, and suggest any modifications that would be recommended to control leakage or seepage pressures.

(d) If the reservoir is 10 km wide and a maximum steady wind velocity of 80 km h^{-1} is assumed, specify the freeboard and rip-rap requirements.

Problem 2 Figure 5.33 shows a dam design section and lists the material properties proposed for a 50 m high earth dam. The dam is proposed for a location where the foundation is of deep alluvial materials of generally low permeability ($k \leq 10^{-8}$ m s^{-1}), but where the expected centerline settlement is estimated to be close to 4 m, with about 2 m of settlement developing after the dam had been completed.

(a) Describe the features of the design that are particularly adapted to perform satisfactorily under the high expected settlements.

(b) What construction problems might be anticipated if the materials are compacted according to specifications? Indicate any changes that you would recommend and the reason for these.

(c) Evaluate the suitability of the materials selected with respect to seepage considerations. Indicate any changes that you would recommend and the reason for these.

(d) Evaluate the distortion in the impervious and semi-pervious zones, and recommend testing procedures to evaluate the performance of these materials under these expected distortions.

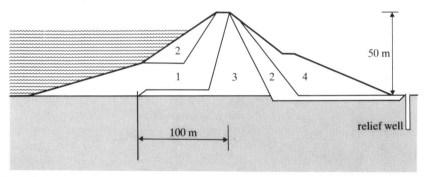

Figure 5.33 Dam section and materials.

Material property	Zone 1: impervious	Zone 2: select	Zone 3: semi-pervious	Zone 4: pervious
D_{15} (mm)	0.0053	0.06	0.01	0.2
D_{50} (mm)	0.033	0.43	0.08	4.5
D_{85} (mm)	0.08	3.8	0.45	25
w_L (%)	17.6	NA	NA	NA
w_P (%)	4.3	NA	NA	NA
w_{opt} (%)	12.8	9.5	8.5	7.0
ρ_d (max) (kg m^{-3})	2050	2020	1920	2160
c' (kPa)	10	0	0	0
ϕ' (deg)	32	34	37	38
k (m s^{-1})	5×10^{-8}	10^{-4}	6×10^{-5}	4×10^{-4}
m_v kPa^{-1}	0.00018	0.00012	0.0001	0.00005
ρ_d (kg m^{-3}) ⎱ as	1950	2050	1920	2150
w (%) ⎰ compacted	14.0	9.5	8.5	5.0

216

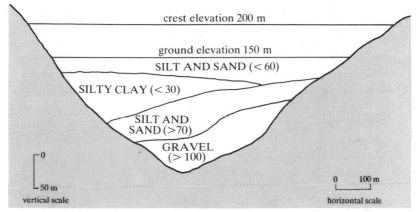

numbers in brackets denote standard penetration blow counts

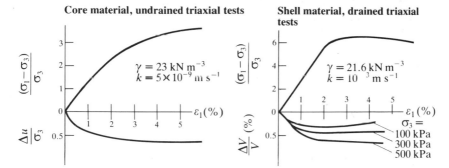

Figure 5.34 Settlement and dam distortion.

(e) Evaluate the risk of hydraulic fracture in the impervious and semi-pervious zones.

(f) Evaluate the factor of safety of the upstream slope if subjected to a rapid lowering of the reservoir level to just below the upstream zone 2 material.

Problem 3   Figure 5.34 shows a section of a valley where a high dam to elevation 200 m is proposed. Triaxial test data for available core and shell materials are shown in a dimensionless form on Figure 5.34. The foundation materials have the following properties:

	$k(\text{m s}^{-1})$	m_v (kPa^{-1})	C_c
silty clay	2×10^{-7}	3.5×10^{-4}	0.38
silt and sand	5×10^{-4}	5×10^{-6}	—
sand and gravel	3×10^{-2}	1×10^{-7}	—

217

(a) Calculate the stable design slope angles for $F = 1.3$, and estimate the differential settlements due to foundation compressibility for a dam founded at elevation 150 m.
(b) Discuss the various foundation treatment methods applicable to this site, and consider the economics as well as the performance of alternative designs. Make recommendations for the most favorable design including slope angles, core and cutoff types and positions and filter requirements.
(c) Show on your design section the locations of recommended instrumentation, and indicate the purpose of each in evaluating the performance.

Problem 4
(a) Explain how a natural saddle as seen on Figure 1.20 can form in a deep lacustrine landform overlying a major rock fault zone.
(b) What type of emergency spillway provisions are there for the two impoundment dams seen on Figure 1.10?

Problem 5 Examine the equations for filter thickness that are given on Figure 5.10, and list the assumptions that are inherent in the equation $t = (Q_c L/k_f)^{1/2}$.

What value of N_f/N_d yields the result that the thickness of the filter is $1.5h(k_c/k_f)^{1/2}$? Discuss the applicability of these equations to standard core thickness recommendations.

6 Ground subsidence and mine backfill

Extraction of mineral resources from the ground has caused surface subsidence and caving in many areas. Past caving problems due to underground mining have sponsored more responsible regulations, but the recognition that base minerals are depleting resources has produced a demand for higher extraction ratios with less ore left underground to support mine openings. Waste materials (waste rock and mine tailings) have been employed in underground cut-and-fill operations for centuries to provide a working floor, but it is only in the last two decades that the real potential of mine backfill in providing ground control is being realized. The use of engineered backfills in a stope-and-pillar type of operation has made it possible to approach full extraction safely with economical high production mining methods while controlling immediate and future subsidence.

6.1 Ground control using backfill

Figure 6.1 shows a typical upper level of an underground room-and-pillar operation. Each mining level is developed as a distinct mining block with drilling being done from the top or the base haulage level. The height of a level (H) is sometimes controlled by the thickness of ore lenses (with waste rock forming sills between levels), or may be restricted by drilling or other physical limitations in deep massive ore bodies. The extraction of stopes is called primary mining and may be initiated on several levels (usually where high ore values are within the reach of early development openings) without creating any problems in ground control or subsidence, because the ore pillars provide the required support. The design dimensions of stopes and pillars will depend on the rock quality, the geometry of the ore body and various economic considerations: good-quality rock may allow in-stope mucking with little bolting requirements and long stopes can be opened up; poor-quality rock excludes working in stopes, and plan dimensions would be limited by the physical requirements for drawpoint cones to allow ore to be drawn from the stope (after blasting, the stope becomes a large bin full of broken rock). The stope length (L) in steeply dipping ore bodies may extend from hanging wall (HW) to foot wall (FW) and the strike width of the stope would then be limited to the safe span of the back. In shallow dipping ore

219

Plan dimensions of ore body at some level below surface

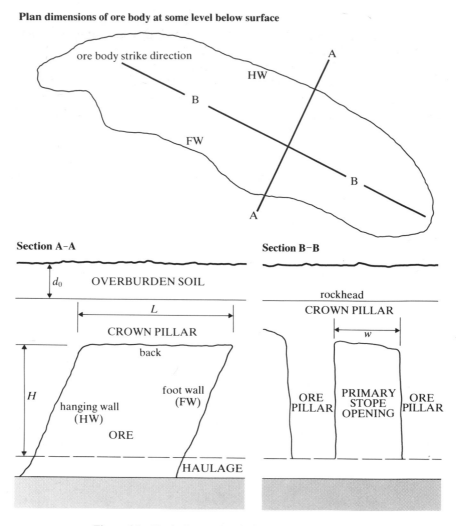

Figure 6.1   Typical upper level of an underground mine.

bodies the HW may form the sloped back, and transverse haulage pillars would be used to subdivide the ore body. Stope spans can also be limited by weaknesses in the hanging wall. In any event, large stope openings with a great variety of heights, lengths and widths are created by high production mining methods. Pillar widths are usually designed for ground control but are sufficiently wide to allow efficient single unit extraction during secondary mining.

Mine backfill is used to provide ground control during pillar extraction operations: primary stopes are backfilled with suitably stable material to prevent caving and resist large closure movements as pillars are sequentially extracted. Since the backfill is never as strong or rigid as the ore, pillar extraction must be properly sequenced to avoid high surcharge or closure stresses in backfill immediately adjacent to the current extraction. As pillar extraction retreats from a particular mining block or ore zone, the pillar stresses will be transferred to the confined backfill and subsidence will develop. The backfill serves two main purposes: to provide temporary ground control as individual pillars are removed and to limit long-term subsidence. Temporary ground control is provided by virtue of the strength and self-weight of a backfill which resists wall sloughing and basal heave in stopes. Resistance to subsidence is provided by virtue of the confined modulus of a backfill, which allows large field stresses to be transferred to the backfill with a limited amount of closure.

6.1.1 Mine backfill materials

Mine wastes are an obvious economic choice for use as mine backfills, although quarried rock and natural borrow materials have been used in cases where suitable mine wastes are not available in sufficient quantities. When operations go underground from an open pit there is often sufficient waste rock which may be efficiently transported and placed in stopes. In

Table 6.1 Mine waste backfills.

Design considerations	Type of waste material	
	Waste rock	Classified tailings
transport to stopes	Dry transport by scoop tram, conveyor, dumped down drilled raises. Relatively expensive	Hydraulic transport in pipelines, poured backfill. Relatively inexpensive
stabilization	Premixing with cement slurry or slurry grouting. Relatively expensive	Cement added into pipeline. Only additional cost is the cost of cement
compressibility	Relatively incompressible if good-quality rock. Poor-quality rock can be very compressible	Same as loose sands but compressibility reduced by stabilization
liquefaction potential	No risk of liquefaction and flow	Careful design of drainage required to avoid risk of liquefaction
drainage requirements	No drainage required	Decant systems are usually required if cement stabilized. Uncemented usually gravity drained
bulkhead requirements	Bulkhead only required to resist slurry grout if grout stabilization used	Careful design of drainage allows timber or concrete block bulkheads to be used

221

deep underground operations the development rock provides only a small portion of the required backfill, and mill tailings are generally delivered underground by pipeline to be used as the main source of fill. Since the porosity of tailings backfills is normally about 0.4, a backfill quantity of up to 0.6 tonnes for each tonne of ore mined is required to fill all mined openings sequentially. Mill tailings are cyclone-separated into classified mine backfill (coarser sands and silts) and slimes (generally less than 0.02 mm) which are wasted to surface impoundments.

The relative advantages and disadvantages of waste rock and classified tailings backfills are listed on Table 6.1. Waste rockfill transport can be prohibitively expensive in deep mines, and waste rock is difficult to stabilize such that it will be self-supporting when adjacent pillars are removed. Filling with waste rock while the pillar ore is being drawn out has been tried, but dilution (drawing of backfill with the ore) can be a costly problem with this method. The major problems with classified tailings backfill are drainage and cement stabilization costs. Recent research efforts have indicated that these problems may be substantially reduced by improvements in placement and design procedures (Wayment 1978, Mitchell et al. 1982). There is little doubt that mine tailings will remain the major backfill material for underground metalliferous mines.

6.1.2 Pouring and drainage of tailings backfills

Classified tailings are normally delivered to underground mine openings by hydraulic transport through boreholes and pipelines. All stope openings must be sealed with bulkheads which are designed to resist the maximum expected backfill and fluid pressures during filling operations. The early use of bulk hydraulic tailings pours was restricted by the consideration that bulkheads would be required to resist a full hydrostatic water pressure given by $u = H\gamma_w$, and many bulkheads were designed to resist a lateral pressure of

$$\sigma_h = H\gamma \qquad (6.1)$$

Massive reinforced concrete bulkheads in the order of 1 m in thickness were used to seal openings typically 3 m × 4 m, and the time required to construct and cure these bulkheads often created unacceptable delays in production. Field measurements (Mitchell et al. 1975, Askew et al. 1978) have shown that bulkhead pressures can be reduced substantially by providing base drainage in stopes and by the arching action which develops in the drained backfill. Provided that drainage is maintained such that the backfill is not subject to spontaneous liquefaction from vibrations or rapid compressive strains, timber bulkheads or concrete block bulkheads are now recommended and may be designed for total pressures of

$$\sigma_h = K_a H\gamma + u \qquad (6.2)$$

in cases where the stope strike length (L) is greater than the stope height (H). In cases where the stope height is greater than the strike length, the fill pressures will be reduced by arching and the bulkheads may be designed for a total pressure of

$$\sigma_\mathrm{h} = \frac{L\gamma}{4 \tan \phi'} + u \qquad (6.3)$$

Smith and Mitchell (1982) provide detailed considerations with regard to bulkhead design and recommend that fill pressures lower than those given by Equations 6.2 and 6.3 can be used in design due to further arching in the drawpoint cone and drift end. Fill pressures on bulkheads are considered to be in the order of 100 kPa and the value of the water pressure u in Equations 6.2 and 6.3 should be evaluated from drainage flow considerations and should be confirmed by monitoring water pressures at piezometer tips located inside the drift plug as shown on Figure 6.2.

Mining engineers generally use percolation rate measurements from a standard percolation tube test to determine the drainage capabilities of a backfill, but Mitchell and Smith (1981) indicate that the measurement of water flow through confined specimens may provide more accurate data. The percolation rate (usually expressed as P in m h^{-1}) is essentially equal to the permeability of the fill material as measured under a gravity gradient close to unity. Percolation tube test results may be corrected for in-place void ratios using the empirical relation.

$$P_\mathrm{corrected} = P_\mathrm{measured} \left(\frac{e_{in\ situ}}{e_\mathrm{measured}} \right)^2 \qquad (6.4)$$

Since classified tailings in a percolation tube are saturated uniform fine sands, the tube percolation value (P_measured) is generally in close agreement with the empirical relation.

$$P_\mathrm{measured} \simeq 50D_{10}^2 \ \mathrm{m\ h^{-1}} \qquad (6.5)$$

where D_{10} is the effective size (in mm).

To provide adequate base drainage during stope pours, the permeability of the stope fill must increase where the stope narrows to form drawpoint cones as shown on Figure 6.2. In particular

$$P_\mathrm{d} \geq P_\mathrm{corrected} \left(\frac{A_\mathrm{s}}{A_\mathrm{d}} \right) \qquad (6.6)$$

where P_d is the percolation rate of drift plug material, A_d the total bulkhead drainage areas (drift sectional areas) and A_s the average stope sectional

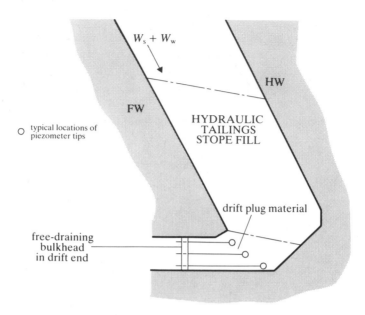

Figure 6.2 Bulkhead drainage.

area. Considering the fact that the drift plug material must be an effective filter (must prevent particle migration and clogging) as well as an efficient filter, it is recommended that the ratio $A_s : A_d$ should not be greater than about 25.

From gravity flow considerations it may be shown that the stope backfill will be free-draining (no water will be impounded on the surface) if

$$P_{\text{corrected}} \geq \frac{W_s \left\{ \left[\dfrac{(1-D)}{D} \right] G_s(1-n) - n \right\}}{G_s \gamma_w (1-n) A_s} \qquad (6.7)$$

where W_s is the weight (kN or tonnes) of solids delivered per hour, D the slurry pulp density, $D = W_s/(W_s + W_w)$, W_w the weight (kN or tonnes) of water per hour in slurry, G_s the specific gravity of solids, n the average porosity of fill in stope, A_s the average sectional area (m²) of stope and γ_w the unit weight (or density) of water, $\gamma_w = 9.81$ kN m⁻³ (1 tonne m⁻³).

In order to ensure that the stope fill is not saturated during pouring operations, it is recommended that the design percolation rate be greater by a factor of 1.5 to 2 than that given by Equation 6.7. Equation 6.7 shows that it is essential to maintain a high pulp density (values between 0.65 and 0.70 can be easily achieved in pipeline transport for specific gravities of 2.8 to 3.5 respectively) if the required quantity of backfill is to be delivered without drainage problems. When hydraulically delivered classified tailings are not

224

sufficiently pervious to maintain free gravity drainage, additional drainage in the form of decant towers have been used. Decant towers must be correctly installed and monitored to ensure that they do not result in increased water pressures in the vicinity of the bulkheads. The use of decant towers and water balance monitoring is discussed in some detail by Smith and Mitchell (1982).

Bulk tailings pours are subject to government regulations in many areas, and one such regulation requires that a cemented tailings plug be poured in stopes not fitted with massive concrete bulkheads. The addition of cement normally reduces the permeability of classified tailings backfill below that required for gravity drainage, and drainage facilities must be provided for uncemented fill placed above the plug. The arrangement shown on Figure 6.3, whereby well points connected to a centrifugal pump are used to evacuate drainage water from a stope, is designed for use with a cemented plug. This arrangement can provide drainage even when the percolation rate is less than that required by Equation 6.7, and eliminates the risk of bulkhead failure even in the extreme event of fill liquefaction. The costs of the well points and pumping systems of Figure 6.3 will be partially recovered by reduced costs in the filter materials and monitoring (pressure and water balance measurements) required with systems such as shown on Figure 6.2. In cases where good-quality development waste rock is available, this rock may be used in place of the cemented tailings plug to eliminate the liquefaction flow risk. The waste rock should be provided with a suitable filter to prevent impregnation by tailings and to allow gravity drainage to occur through the waste rock to the drainage bulkheads.

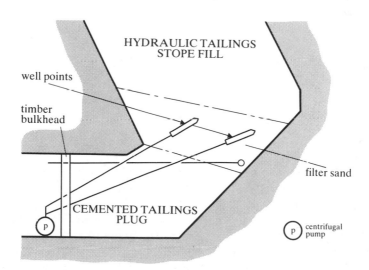

Figure 6.3 Forced drainage with cemented plug.

6.2   Cemented tailings backfill design

Primary stope openings are filled with cemented classified tailings backfill in order to prevent dilution when the pillar ore is removed during secondary mining operations, as shown on Figure 6.4. Cement is added in the hydraulic delivery system and decant pipes are usually required to maintain a surface

(a) **Longitudinal section showing cemented fill being exposed as pillar ore is drawn out**

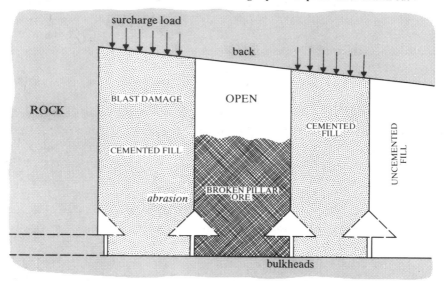

(b) **Forces acting on exposed cemented backfill**

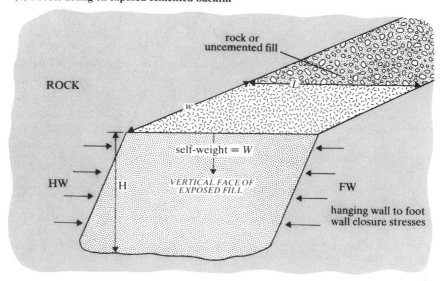

Figure 6.4   Use of cemented tailings backfill.

dry condition during filling. Mass cement additions of between 3.3% (30 : 1 T : C) and 5% (20 : 1 T : C) are common and are always sufficient to eliminate any risk of bulkhead failure or fill liquefaction. During pillar retreat the local back is supported by surrounding rock masses and closure strains are not generally sufficient to cause failure in the fill. The fill is, therefore, designed to be a free-standing structure over the height of the face which is exposed as the pillar ore is removed. As the retreat of pillars progresses (in an orderly sequence), all of the ground stresses will come into equilibrium with the backfill.

While adjacent blasting is likely to cause some ore dilution from cemented backfill, the typical fill appears to be capable of absorbing the energy from controlled blasting operations without mass failure.

6.2.1 Strength of cemented tailings backfills

Figure 6.5 shows typical results from triaxial tests on 28 day cured cemented tailings samples. The overall strength can be approximated by c' and ϕ' as shown on Figure 6.5 or by a low-stress cement bond strength (C_b) which can be defined by unconfined compression test results. At low cement contents (less than 5% cement by dry mass) the stress–strain curve is usually non-brittle, but brittleness increases with cement content and decreases with confining pressure as shown on Figure 6.5. Preliminary design of cement strengths can be carried out using the results of unconfined tests, but selected design mixes should be subjected to triaxial testing and c', ϕ' analysis. The assumption of $r_u = 0$ is usually justified for cured backfills, but bulkhead pressure measurements should be conducted if there is any doubt about the internal drainage of the material. Hydraulic pours are known to cause material segregation as shown on Figure 6.6, but the average strength is generally found to be comparable with laboratory control test results. Figure 6.7 shows that the strength of samples can vary significantly, depending mainly on the mineralogy of the tailings, the pour pulp density and the curing conditions (Weaver & Luka 1970, Thomas *et al.* 1979, Mitchell & Wong 1982). Very low strengths can result from adverse chemical reactions or bacterial action in a highly acidic environment. *In-situ* testing methods are not well developed for backfills, but block sampling can often be done by mining into a fill through a sublevel opening to retrieve samples for trimming and testing. The cost of cementing fill can be a major operational cost in mine backfilling, and field sampling and testing should always be considered to confirm or modify the design, particularly when strength deterioration is noted in laboratory samples. The unconfined compressive strength of a good-quality cemented backfill is expected to increase with curing time according to the form

$$[\sigma_1]_f = A + BC^2 \log t \quad \text{kPa} \tag{6.8}$$

227

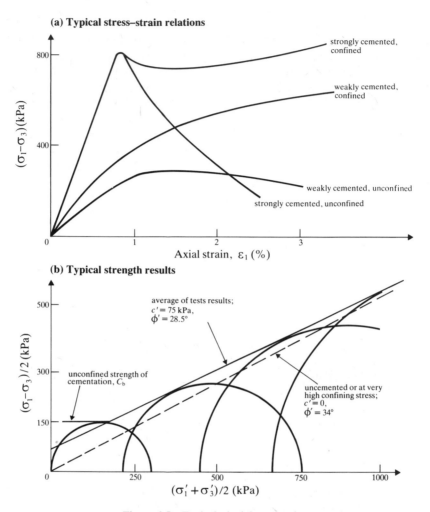

(a) Typical stress–strain relations

strongly cemented, confined

weakly cemented, confined

weakly cemented, unconfined

strongly cemented, unconfined

$(\sigma_1 - \sigma_3)$ (kPa)

Axial strain, ε_1 (%)

(b) Typical strength results

average of tests results;
$c' = 75$ kPa,
$\phi' = 28.5°$

unconfined strength of cementation, C_b

uncemented or at very high confining stress;
$c' = 0$,
$\phi' = 34°$

$(\sigma_1 - \sigma_3)/2$ (kPa)

$(\sigma_1' + \sigma_3')/2$ (kPa)

Figure 6.5 Typical triaxial test results.

where A and B are constants, C is the cement content (%) and t is the curing time in days. Twenty-eight day strengths for $A = 30$, $B = 5$ in Equation 6.8 are shown on Figure 6.7.

Limited statistical testing indicates that sample variation is of the same order expected in natural soils materials and similar safety factors should be considered satisfactory.

6.2.2 Limit design in cemented backfills

Early design of cemented backfills was usually based on a free-standing wall concept which gives a limiting unconfined compressive strength of

228

$[\sigma_1]_f = \gamma H$. Later design shifted to using planar or circular arc analysis which gives $[\sigma_1]_f = 0.5\gamma H$. It is apparent from Figure 6.4 that an exposed backfill is neither a free-standing wall nor a two-dimensional slope. If, for example, the exposed face length (L) becomes small (as in a narrow steeply dipping ore body) the cemented fill could bridge across this length and the cement requirement would be independent of the fill height. The three-dimensional geometry of the fill should be considered in evaluating the strength requirement. It might also be suggested that the cement strength should increase from nearly zero at the top of the fill to a maximum value at the base of the fill. There is, however, no economic gain in this design and there are potential difficulties in operational control. For a constant-strength design the cohesive strength requirement of a given cemented backfill will be a function of the geometrical parameters and the value of ϕ'. When $\phi' = 0$, the unconfined compressive strength requirement is given by

$$[\sigma_1]_f = 2C_b = \gamma HF/N_f \qquad (6.9)$$

where N_f is a stability number to be determined.

Figure 6.6 Segregation in cemented fill.

229

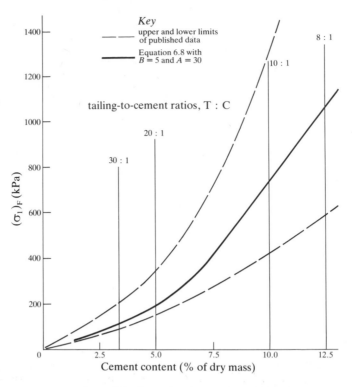

Figure 6.7   Unconfined strength at 28 days.

A generalized failure mechanism for high fills is shown on Figure 6.8, where it is assumed that shear stresses at the wall contacts support some of the self-weight of the sliding block. Assuming that the critical failure angle is given by $\alpha = 45° + \phi'/2$, a net block weight is calculated as

$$W_{N} = wH_{f}(\gamma L - 2c')$$

where $H_{f} = H - \tfrac{1}{2}\, w \tan \alpha$, and the safety factor is obtained as

$$F = \frac{\tan \phi'}{\tan \alpha} + \frac{2c'L}{H_{f}(\gamma L - 2c')\sin 2\alpha} \tag{6.10}$$

The minimum value of ϕ' for tailings sands can be taken as 30°, giving tan $\phi'/\tan \alpha = 0.33$. Thus, for $F \geq 1.33$, the required apparent cohesion is given as

$$c' = \frac{\gamma L H_{f} \sin 2\alpha}{2(L + H_{f} \sin 2\alpha)} \tag{6.11}$$

230

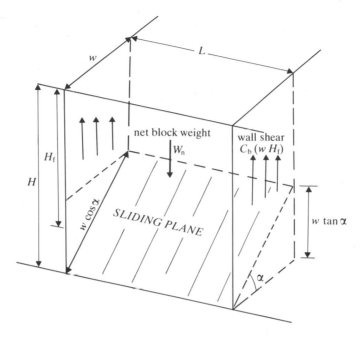

Figure 6.8 Confined block mechanism.

Figure 6.9 Typical failure in model test.

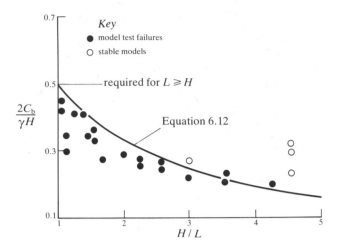

Figure 6.10 Model test results.

In many cases the fill width (w) is small compared to the fill (H) and, for $\phi = 0$, the value of N_f in Equation 6.9 can be evaluated as $(1 + H/L)$, giving

$$[\sigma_1]_f = \gamma HF/(1 + H/L) \tag{6.12}$$

Physical model tests reported by Mitchell *et al.* (1982) showed that failures developed by the mechanism shown on Figure 6.8, and a typical failure is shown on Figure 6.9. Figure 6.10 shows the correlation between model failure data and Equation 6.12. When L is much less than H, Equation 6.12 gives $[\sigma_1]_f = \gamma L$, which is equivalent to developing sufficient wall shear that $W_N = 0$. As the value of H/w increases, the safety factor decreases in Equation 6.10, but Equation 6.12 is theoretically conservative for all values of H/w. For values of H/w greater than about 5, however, it is recommended that the design be checked using retaining wall concepts as discussed in the following section.

6.3 Use of uncemented tailings backfill

During pillar retreat the pillar voids are generally filled using uncemented classified tailings, as indicated on Figure 6.4. Uncemented backfill can also be used for primary stope filling but a remnant pillar must be left to support this fill during pillar extraction. Remnant pillars can represent 10 to 15% of the potentially recoverable ore in a mine but it may be good mine economics to leave this ore if the cost of stabilizing primary fill (cement costs plus decant drainage costs) is greater than the profits to be realized from the remnant pillar ore.

232

An uncemented primary backfill must be free-draining to avoid blast liquefaction (water balance measurements and water content sampling are recommended) and the remnant is designed as a retaining wall for the uncemented fill. As pillar extraction is retreated from an area the remnants are expected to crush and the ground stresses will come into equilibrium with the confined uncemented backfill.

To design the remnant it is necessary to estimate the lateral pressure on this remnant. Terzaghi (1943) calculated that the vertical stress in a dry frictional soil arching between two rigid walls would be

$$\sigma_v = \frac{L\gamma}{2K \tan \phi} \left\{ 1 - \exp[-2K(\tan \phi)z/L] \right\} \qquad \begin{array}{c} (6.13) \\ (\equiv 5.7) \end{array}$$

where L is the distance between walls and K is a constant to be evaluated. A simplified analysis with reference to the diagram on Figure 6.11 where $H > L$ gives

$$\frac{\partial \sigma_v}{\partial z} = \gamma - \frac{2K\sigma_v \tan \phi'}{L}$$

which gives a maximum σ_v as

$$\sigma_{v \max} = L\gamma/2K \tan \phi' \qquad (6.14)$$

where K will be somewhere between K_o and K_p. For $\phi' = 30°$, $K_o = 0.5$ and $K_p = 3.0$. Since the strains developed in a settling backfill are small compared to the strains required to develop full passive pressure, a value of $K = 1$ is recommended. With $K = 1$, the maximum vertical stress is $L\gamma/2 \tan \phi'$, representing a physical arch of height $\frac{1}{2}L \cot \phi'$.

Some numerical analysis predictions have supported the arching concept in uncemented backfills (see, for example, Barrett *et al.* 1978) but more important is the support provided by physical in-stope measurements. Askew *et al.* (1978) used pressure cells to measure backfill pressures in a stope where $L = 45$ m, $\phi' = 36°$ and $\gamma = 20$ kN m^{-3}, and the average measurements are compared to the arching prediction in Figure 6.11, confirming the arching phenomenon and indicating that the lateral pressure is about 50% of the predicted vertical pressure. Thus a design lateral pressure of

$$\sigma_h = L\gamma/4 \tan \phi'$$

would appear to be an appropriate value for the general case of $H \gg L$.

A high remnant ($H > L$) may be considered as a fixed end beam of span L and the design will be governed by the allowable tensile stress in bending, given as

233

$$\sigma_t = wL^2/2t^2$$

where t is the remnant thickness and

$$w = L\gamma/4 \tan \phi'$$

Then

$$\frac{t}{L} = \frac{L\gamma}{8\sigma_t \tan \phi'} \qquad (6.15)$$

where σ_t is usually limited to the compressive pre-stress in the remnant, since the rock will not be capable of supporting tensile stresses. Equation 6.15 is plotted on Figure 6.12 for a range of representative field conditions.

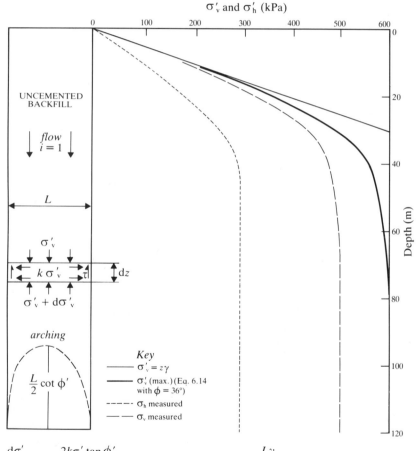

$$\frac{d\sigma'_v}{dz} = \gamma - \frac{2k\sigma'_v \tan \phi'}{L} \; ; \; \text{for } d\sigma'_v = 0, \; \sigma'_v \text{ (max)} = \frac{L\gamma}{2k \tan \phi'}$$

Figure 6.11 Backfill pressures.

234

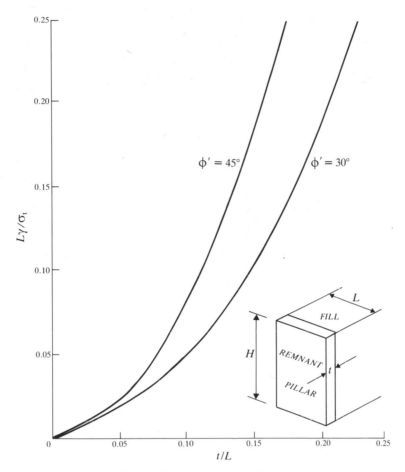

Figure 6.12 Remnant pillar design.

Once the remnant height exceeds its length, the thickness is theoretically independent of the height. In practice, however, the danger of blast fracturing the remnant increases as its height increases, and the ratio H/t should not be allowed to exceed 10. For economical application of this method of backfill support the value of $L\gamma/\sigma_t$ should generally be less than 0.05, and the pillar thickness with suitable safety factor can be approximated as

$$t = 2.5\frac{L^2\gamma}{\sigma_t} \tag{6.16}$$

Since H generally increases with increased L, Equation 6.16 provides an increase in the safety factor as the remnant height increases. For the unusual case where $H \leq L$, the remnant would carry the loads by bending in the vertical plane and H can be substituted for L in the design calculations.

235

It is apparent that only certain mine geometries would be suitable for uncemented fill and remnant pillar operations or, as a corollary, the backfilling costs can be a significant factor in establishing the most economical mining layout: this is true of both cemented and uncemented backfilling, as discussed briefly by Mitchell and Smith (1979) and by Thomas *et al.* (1979).

6.4   Subsidence and surface effects

Traditional subsidence analysis has developed from coal-mining experience where caving developed above the mined-out seam, and an angle of draw accompanied the caving (this angle depending on the characteristics of the overlying materials and varying between 35° and 90° to the horizontal). The maximum subsidence develops directly over the seam and is expressed as $\delta_{max} = \lambda m$ where m is the seam thickness and λ is a subsidence factor that varies between 0.15 and 0.9 depending on the rock quality, depth of cover and whether any waste materials were stowed in the mined openings. Predictions are made mainly on past experience (Brauner 1973). The relationships derived from coal-mining experience are not generally considered applicable to metalliferous underground mining operations where backfilling of all stope and pillar openings is practiced.

6.4.1   Subsidence due to crown pillar deformation

During primary mining operations there should be no inelastic subsidence effects since the pillars are designed to support all roof and wall loadings, and sufficient crown pillar thickness to resist caving is provided. The pillars are often much wider than necessary for ground control in order that they may be extracted by the same methods and with the same efficiency as the primary stopes. With sequential pillar extraction, the local field stresses are shared between existing pillars and the backfill but following pillar retreat the field stresses will come into equilibrium with the backfill. It is during and following the secondary mining stage (pillar extraction) that subsidence will develop.

Subsidence calculations involve both instabilities in the surrounding rock and settlements (or displacements) in the backfill. In recent years a number of finite element analyses have been developed to predict subsidence, but the choice of moduli for both the rock and the backfill appears to be a matter of speculation. Rock ore moduli have been reduced by factors of 2 to 10 to provide a bulk modulus for the rock, and the backfill modulus is virtually ignored with only the self-weight stresses being considered useful in ground control. At the present time these numerical analyses are most useful in providing a detailed overview of the potential stress concentrations during a particular mining sequence, but subsidence magnitudes should also be estimated using approximate solutions outlined below.

236

Long-term subsidence can be estimated using a beam on an elastic foundation approach. The equations, following Hetenyi (1958), can be developed with reference to Figure 6.13 by introducing a backfill subgrade modulus $K = E_c/H_m$, where E_c is the constrained modulus of the fill and H_m is the maximum thickness of the backfill influenced by the loading (q). From bending considerations it may be shown that caving will develop if

$$M > \sigma_h(d_b)^2/6 \qquad (6.17)$$

where

$$M = \frac{\sigma_v}{\lambda^2} \left(\frac{b - a}{c} \right)$$

and

$$
\begin{aligned}
a &= \sinh(\lambda L/2)\cos(\lambda L/2) \\
b &= \cosh(\lambda L/2)\sin(\lambda L/2) \\
c &= \sinh \lambda L + \sin \lambda L \\
\lambda &= (K/4E_r I)^{1/4} \\
I &= d_h^3/12
\end{aligned}
$$

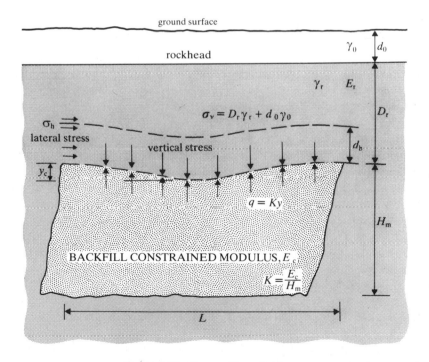

Figure 6.13 Crown pillar subsidence.

and E_r is the assumed bulk modulus of the rock and d_b the assumed depth of intact or bolted rock strata.

Various assumptions regarding E_r and d_b can be used to estimate the probability of caving. Generally larger values of the depth of cover (D_r) provide better quality rock and higher lateral pre-stresses, such that when $D_r \gg L$ extensive caving is not likely to develop and the surface subsidence can be estimated as

$$y_c = \frac{\sigma_v}{K}\left(1 - \frac{2(a+b)}{c}\right) \tag{6.18}$$

When $L \geq D_r$ or the rock quality is poor, caving is likely to develop, and

$$y_c = \frac{\sigma_v}{K} \tag{6.19}$$

can be used to estimate the maximum subsidence for the condition of tight backfilling. Tight backfilling is necessary to provide contact between the backfill and overlying rock (crown pillar) when there is a possibility of caving through to the surface (when $L \geq D$). A bulking in rock volume occurs with caving such that voids below the back may be filled by cave materials which will transfer the weight of the overlying materials to the backfill, but such voids should be kept to less than about 5% of the depth of cover in order to choke off the caving in the lower part of the rock cover. In most cases the rock quality decreases close to rockhead, and substantial increases in groundwater seepage could develop if caving extends close to the surface.

The value of backfill subgrade modulus is calculated as the confined modulus divided by the total depth of backfill that will be compressed. Typical confined test results are shown on Figure 6.14: while cementation will stiffen the backfill response to early loading, the cement bonds will crush under higher loads and the total compression will be nearly independent of the cement content at a stress of 10^4 kPa (equivalent to about 300 m of overlying rock weight). Increased backfill density will greatly increase the resistance to compression: a typical in-place porosity is about $n = 0.43$ and, since $\Delta n = \epsilon_1(1 - n)$, a strain of about 12% would decrease this porosity to 0.38. If the backfill could be densified to $n = 0.38$ before any external load came to bear on it, the early compression would be substantially reduced. Unfortunately, in-place densification of hydraulic fills in open stopes is not considered to be economically viable. From Figure 6.14 it is noted that the stiffness increases as the stress level increases. At any predetermined stress level the confined modulus (E_c) is obtained as $E_c = \Delta\sigma_1/\Delta\epsilon_1$. For stresses where the relationship forms a straight line on the semi-logarithmic plot of Figure 6.14, the average constrained modulus is given as

238

$$E_c = \Delta\sigma/C_\epsilon \log\left(1 + \frac{\Delta\sigma}{\sigma_0}\right) \qquad (6.20)$$

where $\Delta\sigma$ is the stress increase, σ_0 is the initial stress and C_ϵ is defined on Figure 6.14. Subsidence calculations can be carried out for various sections along the strike of the ore body, and ground distortions (both longitudinal and transverse) can be estimated from the results.

Additional subsidence can develop due to overburden consolidation if the groundwater table is lowered. While this can easily be calculated from the results of overburden consolidation tests, two situations tend to predominate when underground mining approaches rockhead: when the land use is minimal and the seepage flows can be economically pumped from underground, the watertightness of the crown pillar is of little concern and the additional subsidence is of less concern; when the surface is developed, the water table must be preserved and there is no additional subsidence. In the latter case caving must be avoided and grouting of the crown pillar may have to be carried out. Figure 6.15 shows an inverted well arrangement to evaluate the mass permeability of a crown pillar underlying a relatively pervious soil aquifer from an underground development drift. The mass permeability may be estimated from the equation

$$k_c \simeq Q/L(d_c + h) \qquad (6.21)$$

where h is the water head above rockhead. The measurement of mass permeability allows calculations of seepage flows and aquifer balance to be made before mining up toward rockhead, and the necessity for grouting can be evaluated.

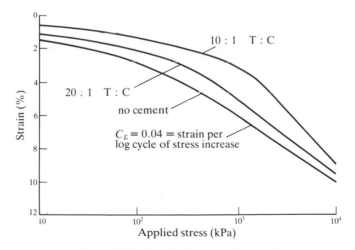

Figure 6.14 Confined compression results.

239

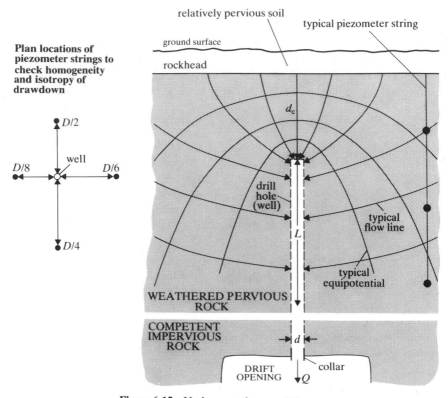

Plan locations of piezometer strings to check homogeneity and isotropy of drawdown

ground surface

rockhead

relatively pervious soil

typical piezometer string

d_c

drill hole (well)

typical flow line

L

typical equipotential

WEATHERED PERVIOUS ROCK

COMPETENT IMPERVIOUS ROCK

d

collar

DRIFT OPENING

Q

$D/2$

well

$D/8$

$D/6$

$D/4$

Figure 6.15 Underground permeability test.

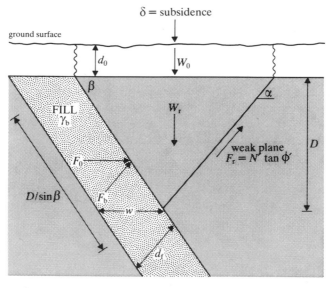

δ = subsidence

ground surface

d_0

W_0

β

α

FILL γ_b

W_r

weak plane $F_r = N' \tan \phi'$

D

F_0

$D/\sin\beta$

F_b

w

d_f

Figure 6.16 Hanging wall slump.

6.4.2 *Subsidence due to hanging wall slump*

If weak planes daylight in the hanging wall of a steeply dipping ore body, long-term slippage on these planes could result in high closure stresses being applied to the backfill, and the compression in the backfill could produce subsidence, as illustrated on Figure 6.16. Using a total backfill resistance derived from resistance due to self-weight (F_0) plus the resistance due to backfill subgrade modulus (F_b), equilibrium would be attained when

$$(W_r + W_0) \sin \alpha = F_r + F_b + F_0 \cos \alpha$$

where

$$
\begin{aligned}
W_r &= D^2 \gamma_r (\cot \alpha + \cot \beta)/2 \\
W_0 &= d_0 \gamma_0 D (\cot \alpha + \cot \beta) \\
F_r &= (W_0 + W_r) \cos \alpha \tan \phi' \\
F_b &= K_c \delta D / \sin \alpha \sin \beta \\
K_c &= E_c / d_f \\
F_0 &= K_0 D^2 \gamma_b / 2 \\
K_0 &= (1 - \sin \phi')
\end{aligned}
$$

The maximum surface subsidence (δ) is then estimated as

$$\delta = \frac{(W_r + W_0)(1 - \tan \phi'/\tan \alpha) - K_0 D^2 \gamma_b (\cot \alpha)/2}{K_c D \sin^2 \alpha \sin \beta} \tag{6.22}$$

If the groundwater table is above the base of the sliding block, the value of W_r in Equation 6.22 should be modified using

$$W_r' = W_r (1 - h_w \gamma_w / D \gamma_r)$$

to obtain an estimate of the subsidence of a partly submerged sliding block.

6.5 Problems on mine backfill and subsidence

Problem 1 If hydraulic delivery of classified sands for backfilling is maintained at a pulp density of $D = G_s/(G_s + 1.5)$, calculate the solid fraction of the slurry by volume. Show that the linear filling rate is given by

$$dH = W_s \, dt / G_s \gamma_w (1 - n) A_s$$

and that the required percolation rate for gravity drainage is given as

$$P \geq (1.5 - 2.5n) \, dH/dt$$

241

Problem 2   If a hydraulic pour of 50 m in height had a percolation rate of 10 cm/h and a porosity of 0.45, how long would be required for the backfill to drain internally after pouring was completed? If $G_s = 3.0$ for this backfill and it is composed of uniform-sized, rounded grains, what residual water content could be tolerated without danger of blast liquefaction?

Problem 3   If a classified tailings sand fill having $D_{10} = 0.037$ mm was used for cut-and-fill mining support, what minimum pour depth would be required in order that the fill surface could be used to support drilling equipment within a day or two of pouring? Consider capillary effects, make an engineering estimate and specify testing which you would recommend to evaluate this problem further.

Problem 4   Using the analyses outlined in this chapter, calculate the most economical stope and pillar dimensions using the following assumptions and ignoring all other possible costs and restrictions:

(a) All mine openings are to be filled with classified tailings by hydraulic delivery. Primary stopes are to be filled with cemented tailings so that pillars may be completely recovered. The strike length (L) of the mining block is 30 m and the block height can range from 30 to 120 m.
(b) Uncemented tailings will be free-draining.
(c) Cemented tailings will require decanting using towers at a cost of $0.2 per m³ of cemented backfill. Cement cost is $40 per tonne of cement used. $F = 1.25$ used in design.
(d) Pillar widths must be at least 30% of the total pillar–stope unit width and 20 m is the minimum width of stopes or pillars.
(e) Pillar height-to-width ratios will not exceed 6 and pillar widths will not exceed 40 m.
(f) Pillar extraction costs per tonne are 1.4 times the primary mining cost, decreasing with volume (V_p) according to cost per tonne = $11.2/\ln V_p$.
(g) The ore mass density is 3.5 tonne m⁻³ and the fill requirement is 0.55 tonnes of backfill for each tonne of ore mined.
(h) Cemented tailings strength is to be designed on 28 day unconfined tests using Equation 6.8 with $A = 30$, $B = 5$.
(i) Primary stoping costs decrease with stoping volume (V_s in m³) according to cost per tonne = $8/\ln V_s$.

If pillar ore produces on average a net profit of $0.3 per tonne, evaluate the economics of using uncemented backfill and remnant pillars (assume a lateral pillar pre-stress of 20 kPa) instead of cemented primary backfill.

Problem 5   Figure 6.17 shows a typical section of a mined-out and backfilled underground mine where only a few mining blocks are still active.

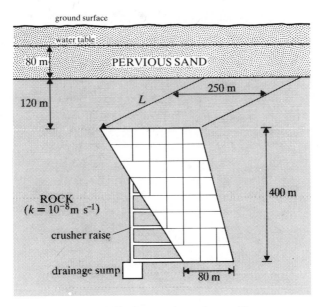

Figure 6.17 Seepage into mine workings.

(a) Estimate the quantity of seepage flow into the mine if $L = 1000$ m.
(b) Estimate the quantity of seepage flow into the mine if subsidence caving increased the average mass permeability of the 120 m crown pillar to 10^{-6} m s^{-1} over 50% of the mine plan area of 2.5×10^5 m^2.

Problem 6 Using the method developed for the mechanism on Figure 6.16 and data from Figure 6.14 calculate the subsidence assuming (a) no cement and (b) 10:1 T:C where the following parameters apply: $w = 10$ m, $\gamma_r = 28$ kN m^{-3}, $\gamma_b = 20$ kN m^{-3}, $\beta = 70°$, $\alpha = 60°$, $d_o = 30$ m, $\phi' = 30°$, $D = 200$ m and $r_u = 0$.

Using the same data estimate the subsidence in the above case by calculating the average stress imposed on the backfill due to rocksliding and considering one-dimensional compression of the backfill under this imposed stress. Compare these answers and comment on any differences.

Problem 7 A proposed underground mine opening is idealized as an infinite slot in a rock of variable permeability, as shown on Figure 6.18. The overburden soil is a normally consolidated silty clay with $k = 10^{-7}$ m s^{-1}, $e_0 = 1.6$ and $C_c = 0.3$ overlying a confined sandy aquifer.

(a) Calculate the maximum potential flow rate into the mine opening.
(b) Calculate the ground subsidence due to de-watering and make recommendations with respect to grouting if the surface water is to be maintained (average infiltration from rainfall is 0.2 m yr^{-1}).

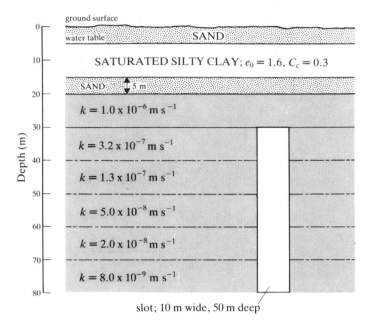

Figure 6.18 Subsidence calculations.

(c) Estimate the surface subsidence due to crown pillar deformation assuming $E_r = 10^9$ kPa, $E_c = 10^6$ kPa, $d_b = 10$ m, $\sigma_h = 10$ MPa and a void of 1.5 m between the crown pillar and the backfill.

Problem 8 List the requirements for dimensional similitude of the model studies depicted on Figures 6.9 and 6.10, and discuss how saturation or drainage of the model pour would affect this similitude.

Problem 9 Given $\sigma_v = 1365$ kPa, $\sigma_h = 9650$ kPa, $E_r = 3.4 \times 10^7$ kPa, $L = 79.3$ m, $K = 2.9 \times 10^3$ kN m^{-3}, calculate and plot expected subsidence against beam depth for the case of tight backfilling.

Appendix: units and symbols

Metric SI units are used throughout the text and the following conversions are provided:

Measurement	Letter symbol	Common SI unit	Equivalent Imperial unit
length	L	m (metre)	3.28 ft
mass	M	kg (1000 grams)	0.0685 slug ($\equiv$ 2.2046 lb)
force	F	kN (1000 newtons)	224.82 lb
volume (liquid)	V_L	l (litre)	0.220 gallon (0.2642 US gallon)
gravitational acceleration	g	9.81 m s^{-2}	32.2 ft s^{-2}
unit weight	$\gamma = \dfrac{Mg}{V}$	$\dfrac{kN}{m^3} - \dfrac{kg(m\,s^{-2})}{m^3}$	6.366 lb ft^{-3}
unit weight of water	γ_w	9.81 kN m^{-3}	62.45 lb ft^{-3}
density of water	ρ_w	1 tonne m^{-3} (1 g cm^{-3})	1.94 slug ft^{-3}

A area
A_f pore pressure coefficient
a base distance
a_s seepage distance
a_v coefficient of compressibility
B width
$\overline{B}$ pore pressure coefficient
C constant, strength
C_b cement bond strength
C_c compressibility index
C_s swelling index
C_v coefficient of consolidation
C_α creep index
Cu undrained shear strength
c apparent cohesion
c' effective apparent cohesion
D pulp density
D depth factor, diameter, thickness
D_{10} effective grain size
d diameter
d' failure envelope intercept
E modulus of elasticity
E efficiency of cutoff
e void ratio
F factor of safety
F_x force acting in x direction
f force ratio factor
H thickness of stratum
H_c critical center distance

H_d drainage path length
ΔH settlement
h total head
h_c height of capillary rise
h_w height of water (piezometric head)
Δh head loss (potential drop)
i hydraulic gradient
i_c critical hydraulic gradient
K subgrade modulus
K stress ratio
K_a active pressure coefficient
K_p passive pressure coefficient
K_s coefficient of subgrade reaction
K_0 coefficient of earth pressure at rest
k coefficient of permeability
L length
L_c critical center distance
LI liquidity index
m coefficient in stability analysis
m_v coefficient of volume compressibility
N number
N_c bearing capacity factor
N_d number of equipotential drops
N_f number of flow channels
N_0 stability factor
N_s stability factor
n porosity
P force created by a pressure distribution

P	percolation rate	z	depth
P_a	active force	z_w	depth of water
P_p	passive force		
PI	plasticity index	α	angle of failure plane
p_c'	preconsolidation pressure	β	slope angle
p_s	seepage pressure	γ	unit weight
Q	flow quantity (volume per unit time)	γ'	submerged unit weight $(\gamma - \gamma_w)$
q	uniformly distributed stress	γ_d	dry unit weight
R	radius of influence of a well	γ_w	unit weight of water
r	radius	Δ	increment
S_r	degree of saturation	δ	subsidence
St	sensitivity of undrained strength	ϵ_x	strain in the x direction
T_v	time factor	$\bar{\epsilon}$	distortional strain
t	time	ν	Poisson's ratio
U	uplift force	ρ	density
$\overline{U}$	average degree of consolidation	ρ_d	dry density
u	pore-water pressure	ρ_w	density of water
u_x	pore-water pressure at point x	σ	stress
V	volume	σ'	effective stress
v	velocity of flow	$[\sigma_1]_f$	unconfined compressive strength
W	weight	τ	shearing resistance
W'	effective weight (weight minus buoyancy)	τ_f	shearing strength on a failure surface
		τ_m	mobilized shearing resistance
w	water content	ϕ	apparent angle of frictional shearing resistance
w_L	liquid limit		
w_P	plastic limit	ϕ'	effective apparent angle of shearing resistance
w_S	shrinkage limit		
x	co-ordinate distance	ϕ_r'	residual angle of shearing resistance
y	co-ordinate distance	ψ'	angle of failure envelope

References

Aberfan Tribunal Report 1968. A selection of technical reports submitted to the Aberfan Tribunal, Welsh Office, London. London: HM Stationery Office.

Algamor, G. and G. Wiseman 1977. Analysis of submarine slumping in the continental slope off the southern coast of Israel. *Marine Geotech.* **2**, 349–88.

Anderson, D. and K. W. Brown 1981. Organic leachate effects on the permeability of clay liners. In *Land disposal: hazardous waste*, Proc. 7th Ann. Res. Symp., 119–30. Washington, DC: US Environmental Protection Agency.

ASCE 1974a. *Subsurface investigation for underground excavation and heavy construction.* Proc. Engng Foundation Conf., New Hampshire, August 1974. New York: Am. Soc. Civ. Engrs.

ASCE 1974b. *Inspection, maintenance and rehabilitation of old dams.* Proc. Engng Foundation Conf., Pacific Grove, CA, Sept. 1973. New York: Am. Soc. Civ. Engrs.

ASCE 1978. *Soil improvement: history, capabilities, outlook.* Rep. Cttee on placement and improvement of soils, ASCE Geotech. Engng Div.

Askew, J. E., P. L. McCarthy and D. J. Fitzgerald 1978. Backfill research for pillar extraction at ZC/NBHC. In *Mining with backfill*, Proc. 12th Can. Rock Mech. Symp., Sudbury, 100–10. CIM Spec. Vol. 19. Montreal: Can. Inst. Mining Metall.

Attewell, P. B. 1977. Large ground movement and structural damage caused by tunnelling below the water table in a silty alluvial clay. In *Large ground movements and structures*, J. D. Geddes (ed.). New York: Wiley.

Avery, T. E. 1968. *Interpretation of aerial photographs*, 2nd edn. Minnesota: Burgess.

Avgherinos, P. J. and A. N. Schofield 1969. Drawdown failures of centrifuge models. *Proc. 7th Int. Conf. on soil mechanics and foundation engineering*, Mexico, vol. 2, 497–506. Mexico: Sociedad Mexicana de Mecánica de Suelos.

Azzouz, A. S., R. J. Krizek and R. B. Corotis 1976. Regression analysis of soil compressibility. *Soils Foundns (Tokyo)* **16**(2), 19–29.

Barden, L. 1965. Consolidation of clay with non-linear viscosity. *Geotechnique* **15**(4), 345–62.

Barrett, J. R. and P. J. Moore 1975. Design criteria for rapid drawdown in earth–rock dams. In *Numerical analysis of dams*, O. J. Naylor, K. G. Stagg and O. C. Zienkiewicz (eds), Proc. Int. Symp., Swansea, 809–28. University College of Wales, Swansea.

Barrett, J. R., M. A. Coulthard and P. M. Dight 1978. Determination of fill stability. In *Mining with backfill*, Proc. 12th Can. Rock Mech. Symp., Sudbury, 85–91. CIM Spec. Vol. 19. Montreal: Can. Inst. Mining Metall.

Barton, N., R. Lien and J. Lunde 1974. Engineering classification of rock masses for the design of tunnel support. *Rock Mech.* **6**(4), 189–236.

Basham, P. W., D. H. Weichert and M. J. Berry 1979. Regional assessment of seismic risk in Eastern Canada. *Bull. Seism. Soc. Am.* **69**(5), 1567–602.

Basore, C. E. and J. D. Boitano 1969. Sand densification by piles and vibroflotation. *ASCE J. Soil Mech. Foundn Engng* **95**(6), 1303–23.

Baumann, V. and G. E. A. Bauer 1974. The performance of foundations on various soils stabilized by the vibro-compaction method. *Can. Geotech. J.* **11**, 509–53.

Bazett, D. J. 1961. Field measurements of pore water pressures. *Proc. 12th Canadian Soil Mechanics Conf.*, Ottawa, 2–15. Ottawa: National Research Council of Canada.

Bea, R. G. and M. E. Audibert 1980. Offshore platforms and pipelines in Mississippi River delta. *ASCE J. Geotech. Engng* **106**(8), 853–69.

Bernell, L. 1964. Measurements in the Messaure dam, a rockfill structure with wet compacted moraine core. *Proc. 8th Int. Congr. on large dams*, Edinburgh, vol. 2, 317–33. Paris: International Commission on Large Dams.

Bieniawski, Z. T. 1974. Geomechanics classification of rock masses and its application in tunnelling. *Proc. 3rd Int. Cong. on rock mechanics*, Denver, vol. 2, 27–32. Washington, DC: National Academy of Sciences.

Bishop, A. W. 1955. The use of the slip circle in the stability analysis of slopes. *Geotechnique* **5**(1), 7–17.

Bishop, A. W. 1966. Sixth Rankine Lecture: The strength of soils as engineering materials. *Geotechnique* **16**(2), 91–128.

Bishop, A. W. 1973. The stability of tips and spoil heaps. *Q. J. Engng Geol.* **6**(4), 335–76.

Bishop, A. W. and N. R. Morgenstern 1960. Stability coefficients for earth slopes. *Geotechnique* **10**(4), 129–50.

Bishop, A. W., M. F. Kennard and A. D. M. Penman 1961. Pore pressure observations at Selset Dam. *Proc. Conf. on pore water pressure and suctions in soils*, London, 91–102.

Bishop, A. W., D. L. Webb and A. E. Skinner 1965. Triaxial tests on soil at elevated cell pressures. *Proc. 6th Int. Conf. on soil mechanics and foundation engineering*, Montreal, vol. 1, 170–4. Toronto: University of Toronto Press.

Bjerrum, L. 1967a. Engineering geology of normally consolidated marine clays as related to the settlement of buildings. *Geotechnique* **17**(2), 83–117.

Bjerrum, L. 1967b. Progressive failure in slopes of overconsolidated plastic clays and clay shales. *ASCE J. Soil Mech. Foundn Engng* **93**(5), 3–49.

Bjerrum, L. 1972. Embankments on soft ground. *ASCE Spec. Conf. on earth structures*, Purdue, Lafayette, IN, vol. 2, 1–54. New York: Am. Soc. Civ. Engrs.

Bjerrum, L., J. Løken, S. Heiberg and R. Foster 1971. A field study of factors responsible for quick clay slides. *Norwegian Geotechnic Institute Publications* **85**, 17–26.

Blondeau, F. and D. Queyroi 1976. Rupture d'une fouille experimental dams les alluvions recentes fortement plastiques. *Proc. 6th Eur. Conf. on soil mechanics and foundation engineering*, Vienna. Vienna: Gistel.

Bowles, J. E. 1978. *Engineering properties of soils and their measurement*, 2nd edn. New York: McGraw-Hill.

Bozozuk, M. 1969. A fluid settlement gauge. *Can. Geotech. J.* **6**, 362.

Bozozuk, M. 1972. Downdrag measurements on a 160 ft floating pipe test pile in marine clay. *Can. Geotech. J.* **9**(2), 127–36.

Bozozuk, M. and G. A. Leonards 1972. The Gloucester test fill. *ASCE Spec. Conf. on earth structures*, Purdue, Lafayette, IN, vol. 1, 299–318. New York: Am. Soc. Civ. Engrs.

Brauner, G. 1973. *Subsidence due to underground mining*. Rep. 8571, US Bureau of Mines, Washington, DC.

Broms, B. 1978. Translatory slips in soft clay. In *Evaluation and prediction of subsidence*, S. K. Saxena (ed.), Int. Subsidence Conf., Florida, 169–81. New York: Am. Soc. Civ. Engrs.

Broms, B. B. and H. Bennermark 1967. Stability of clay at vertical openings. *ASCE J. Soil Mech. Foundn Engng* **93**(1), 71–94.

Brown, S. F., A. K. Lashine and A. F. Hyde 1975. Repeated load triaxial testing of a silty clay. *Geotechnique* **25**(1), 95–114.

Burke, H. H. and S. S. Smucha 1981. Geodrain installation at Lornex tailings dam. *Proc. 10th Int. Conf. on soil mechanics and foundation engineering*, Stockholm, vol. 3, 599–602. Stockholm: Liber Tryck.

Burland, J. B. 1971. A method of estimating pore water pressures and displacements beneath embankments on soft natural clay deposits. *Stress–strain behaviour of soils*, Proc. Roscoe Memorial Symp., 505–36. Oxfordshire: Foulis.

CANMET 1972. *Tentative design guide for mine waste embankments in Canada*. Energy, Mines and Resources, Canada, Tech. Bull. TB 145.

Carlyle, W. J. 1969. Llyn Brianne dam, S. Wales. *Civ. Engng Public Works Rev.* **64**(761), 1195–200.

Carson, M. A. 1977. On the retrogression of landslides in sensitive muddy sediments. *Can. Geotech. J.* **14**(4), 582–602.

Casagrande, A. 1940. Seepage through dams. In *Contributions to soil mechanics, 1925–1940*. Boston: Boston Society of Civil Engineers.

Casagrande, A. 1961. First Rankine Lecture: Control of seepage through foundations and abutments of dams. *Geotechnique* **11**(3), 159–81.

Casagrande, A. 1969. Treatment of pervious foundations of great depth. *Proc. 7th Int. Conf. on soil mechanics and foundation engineering*, Mexico, Discussion of session 3. Mexico: Sociedad Mexicana de Mecánica de Suelos.

Cedergren, H. R. 1973. Seepage control in earth dams. In *Embankment-dam engineering – Casagrande volume*, R. C. Hirschfeld and S. J. Poulos (eds), 21–45. New York: Wiley.

Clevenger, W. A. 1956. Experiences with loess as foundation material. *ASCE J. Soil Mech. Foundn Engng* **82**(3), 1025–6.

CMFE 1975. *Canadian manual on foundation engineering*. Montreal: Canadian Geotechnical Society, Engineering Institute of Canada.

Collins, W. E. 1964. Design of Blue Mesa Dam. *Proc. 8th Int. Congr. on large dams*, Edinburgh, vol. 3, 731–44. Paris: International Commission on Large Dams.

Conlin, B. H. 1982. Design considerations for earthquake-resistant mine tailings impoundment structures. *Proc. 34th Canadian Geotechnical Conf.*, Fredericton, NB. Montreal: Canadian Geotechnical Society, Engineering Institute of Canada.

Cooke, J. B. 1960. Discussion: Goschenenalp Dam – Switzerland. *Trans Am. Soc. Civ. Engrs* **125**(2), 575–6.

Crooks, J. H. A., E. L. Matyas and H. M. McKay 1980. Excavation slope stability related to pore-water pressure variations during piling. *Can. Geotech. J.* **17**(2), 225–35.

Cruden, D. M. and J. Krahn 1973. A re-examination of the geology of the Frank slide. *Can. Geotech. J.* **10**(4), 581–91.

Cundall, P. A. 1974. *A computer model for rock-mass behavior using interactive graphics for the input and output of geometrical data*. Report on contract DACW 45-74-C-006, US Army Corps of Engineers, University of Minnesota.

D'Appolonia, D. J., T. W. Lambe and H. G. Poulos 1971. Evaluation of pore pressure beneath an embankment. *ASCE J. Soil Mech. Foundn Engng* **97**(6), 881–97.

Dascal, O. 1979. Hydraulic efficiency of the Manicouagan-3 cutoff. *Can. Geotech. J.* **16**(2), 351–62.

Dascal, O., J. P. Tournier, F. Tavenas and P. LaRochelle 1972. Failure of a test embankment on sensitive clay. *ASCE Spec. Conf. on earth structures*, Purdue, Lafayette, IN, vol. 1, 129–58. New York: Am. Soc. Civ. Engrs.

Davies, W. E. 1973. Buffalo Creek Dam disaster. *ASCE Civ. Engng*, July, 69–72.

Davis, E. H. and G. P. Raymond 1965. A non-linear theory of consolidation. *Geotechnique* **15**(2), 161–73.

Debidin, F. and C. F. Lee 1980. Groundwater and drawdown in a large earth excavation. *Can. Geotech. J.* **17**, 185–202.

Deere, D. U. 1964. Technical description of rock cores for engineering purposes. *Rock Mech. Engng Geol.* **1**(1), 17–22.

Deere, D. U. and F. D. Patton 1971. Slope stability in residual soils. *Proc. 4th Pan Am. Conf. on soil mechanics and foundation engineering*, Puerto Rica, vol. 1, 87–170. New York: Am. Soc. Civ. Engrs.

Deere, D. U., J. E. Monsees and B. Schmidt 1970. *Design of tunnel liners and support systems*. Transportation Research Board, Washington, DC, Highway Research Record 339.

Dishaw, H. E. 1967. Massive landslides. *Photogramm. Engng* **33**, 603–8.

Dobry, R. and L. Alvarez 1969. Seismic failure of Chilean tailings dams. *ASCE J. Soil Mech. Foundn Engng* **93**(6), 237–60.

Duncan, I. M. and H. B. Seed 1966. Anisotropy and stress reorientation in clays. *ASCE J. Soil Mech. Foundn Engng* **92**(5), 21–50.

Dunlop, P., J. M. Duncan and H. B. Seed 1968. *Finite element analysis of slopes in soil.* Rep. TE 68–3. University of California, Dept of Civil Engineering, Berkeley, CA.

Du Pont, De Nemours and Co. 1958. Fill settlement. *Blaster's handbook*, 14th edn. Wilmington, Del.: Du Pont, De Nemours and Co.

Dusseault, M. B. and N. R. Morgenstern 1978. Shear strength of Athabasca Oil sands. *Can. Geotech. J.* **15**(2), 216–38.

Eden, W. J. and M. Bozozuk 1969. Earth pressures on Ottawa outfall sewer tunnel. *Can. Geotech. J.* **6,** 17–33.

Eden, W. J. and R. J. Mitchell 1970. The mechanics of landslides in Leda Clay. *Can. Geotech. J.* **7**(3), 285–96.

Eden, W. J. and H. B. Poorooshasb 1968. Settlement observations at Kars bridge. *Can. Geotech. J.* **5,** 28–45.

Eden, W. J., E. B. Fletcher and R. J. Mitchell 1971. South Nation River landslide, 16 May 1971. *Can. Geotech. J.* **8**(3), 446–51.

Eide, O. and S. Holmberg 1972. Test fills to failure on the soft Bangkok Clay. *ASCE Spec. Conf. on earth structures*, Purdue, Lafayette, IN, vol. 1, 159–80. New York: Am. Soc. Civ. Engrs.

Eisenstein, Z. and J. V. Simmons 1975. Three dimensional analysis of Mica Dam. In *Numerical analysis of dams*, O. J. Naylor, K. G. Stagg and O. C. Zienkiewicz (eds), Proc. Int. Symp., Swansea, 1052–69. University College of Wales, Swansea.

Eisenstein, Z., A. V. Krishnayya and N. Morgenstern 1972. An analysis of cracking at Duncan Dam. *ASCE Spec. Conf. on earth structures*, Purdue, Lafayette, IN, vol. 1, 765–78. New York: Am. Soc. Civ. Engrs.

Ells, R. W. 1908. *Landslide at Notre-Dame de la Salette.* Dept of Mines, Geological Survey, Canada.

ENR 1960. *Engng News Record*, **164**(April), 25.

Farvolden, N. R. and J. P. Nunan 1970. Hydrogeologic aspects of dewatering at Welland. *Can. Geotech. J.* **7**(2), 194–204.

Finn, W. D. L. 1967. Finite element analysis of seepage through dams. *ASCE J. Soil Mech. Foundn Engng* **93**(6), 41–8.

Folkes, D. J. 1982. Control of contaminant migration by the use of liners. 5th Canadian Geological Colloquium. *Can. Geotech. J.* **19**(3), 320–44.

Fredlund, D. G. 1978. Usage, requirements and features of slope stability computer software (Canada, 1977). *Can. Geotech. J.* **15**(1), 83–95.

Fredlund, D. G. 1979. Appropriate concepts and technology for unsaturated soils. 2nd Canadian Geotechnical Colloquium. *Can. Geotech. J.* **16**(1), 121–39.

Frind, O. E. 1970. Theoretical analysis of aquifer response due to dewatering at Welland. *Can. Geotech. J.* **7**(2), 205–16.

Galbiatti, I. V. 1963. The construction of the ICOS cutoff wall at the Quebec Hydro Manicouagan-5 power development. *ASCE water resources engineering conf.*, Milwaukee. New York: Am. Soc. Civ. Engrs.

Gassler, G. and G. Gudehus 1981. Soil nailing – some aspects of a new technique. *Proc. 10th Int. Conf. on soil mechanics and foundation engineering*, Stockholm, vol. 3, 665–70. Stockholm: Liber Tryck.

Gedney, D. S. and W. G. Weber 1978. Design and construction of soil slopes. *Landslides: analysis and control*, 172–90. Spec. Rep. 176, Transportation Research Board, National Academy of Sciences, Washington, DC.

Gibson, R. E. and N. R. Morgenstern 1962. A note on the stability of cuttings in normally consolidated clays. *Geotechnique* **12**(3), 212–16.

Glover, R. E., H. J. Gibbs and W. W. Daehn 1948. Deformability of earth materials and its effect on the stability of earth dams following a rapid drawdown. *Proc. 2nd Int. Conf. on soil mechanics and foundation engineering*, Rotterdam, vol. 5, 77–80. Haarlem: Keesmaat.

Goel, M. C. and N. C. Das 1981. Construction pore water pressures – case study of Ramganga Dam. *Proc. 10th Int. Conf. on soil mechanics and foundation engineering*, Stockholm, vol. 3, 417–20. Stockholm: Liber Tryck.

Golder, H. Q. and D. J. Bazett 1967. An earth dam built by dumping through water. *Trans. 9th Int. Congr. on large dams*, Istanbul, vol. 4, 369–87. Paris: International Commission on Large Dams.

Gordon, J. L. and D. R. Duguid 1970. Experiences with cracking at Duncan Dam. *Trans. 10th Int. Congr. on large dams*, Montreal, vol. 1, 469–86. Paris: International Commission on Large Dams.

Gouchnour, R. R. and O. B. Andersland 1968. Mechanical properties of a sand–ice system. *ASCE J. Soil Mech. Foundn Engng* **94**(4), 923–50.

Graham, J. 1979. Embankment stability on anisotropic soft clays. *Can. Geotech. J.* **16**, 295–308.

Griffin, R. A. and N. F. Shimp 1978. Attenuation of pollutants in municipal landfill leachate by clay minerals. Ill. St. Geol. Surv., Rep. No. EPA-600/2-78-157, prepared for US Environmental Protection Agency.

Griffith, A. A. 1924. Theory of rupture. *Proc. 1st Int. Congr. on Applied Mathematics*, Delft, 55–63.

Guilford, C. M. and H. C. Chan 1969. Some soil aspects of the Plover Cove marine dam. *Proc. 7th Int. Conf. on soil mechanics and foundation engineering*, Mexico, vol. 2, 291–300. Mexico: Sociedad Mexicana de Mecánica de Suelos.

Haxo, H. E. Jr 1981. Durability of liner materials for hazardous waste disposal facilities. In *Landfill disposal: hazardous waste*, Proc. 7th Annu. Res. Symp., 140–56. Washington, DC: US Environmental Protection Agency.

Hemborg, H. B. and W. O. Keightley 1964. Man-made earthquakes test a dam. *Engng News Record*, April, 48–9.

Henkel, D. J. 1960. The relationships between the effective stresses and water content in saturated clays. *Geotechnique* **10**(2), 41–54.

Henkel, D. J. 1970a. Geotechnical considerations of lateral stresses. *ASCE Spec. Conf. on Earth-retaining structures*, Cornell University, Ithaca, NY. New York: Am. Soc. Civ. Engrs.

Henkel, D. J. 1970b. The role of waves in causing submarine landslides. *Geotechnique* **20**(1), 75–80.

Hetenyi, M. 1958. *Beams on elastic foundations*. Chicago: University of Michigan Press.

Hodge, R. A. and R. A. Freeze 1977. Groundwater flow systems and slope stability. *Can. Geotech. J.* **14**(4), 466–76.

Hoek, E. and J. W. Bray 1974. *Rock slope engineering*. London: Institute of Mining and Metallurgy.

Hudson, R. Y. 1954. Wave forces on breakwaters. Symp. on engineering and water waves. *Trans. Am. Soc. Civ. Engrs* **119**, 653.

Hunter, J. H. and R. L. Schuster 1968. Stability of simple cuttings in normally consolidated clays. *Geotechnique* **18**(3), 372–8.

Hvorslev, M. J. 1960. Physical components of the shear strength of saturated clays. *ASCE Res. Conf. on the shear strength of cohesive soils*, Boulder, CO. New York: Am. Soc. Civ. Engrs.

ICOLD 1973. *Lessons from dam incidents*, 2nd edn. Paris: International Committee on Large Earth Dams.

251

Jacob, C. E. 1950. Flow of ground water. In *Engineering hydraulics*, H. Rouse (ed.), 321–86. New York: Wiley.

Jaeger, J. C. 1971. Eleventh Rankine Lecture: Friction of rocks and stability of rock slopes. *Geotechnique* 21(2), 97–134.

James, J. P. and R. T. Wickham 1970. *The failure and rectification of Morwell no. 2 fire service reservoir*, 20–34. ANCOLD Bull. no. 31.

Janbu, N. 1973. Slope stability computations. In *Embankment-dam engineering*, 47–86. New York: Wiley.

Janes, H. W. 1973. Densification of sand for dry dock by terra probe. *ASCE J. Soil Mech. Foundn Engng* 99(6), 451–70.

Jansen, R. B., G. W. Dukleth, B. B. Gordon, L. B. James and C. E. Shields 1967. Earth movement at Baldwin Hills Reservoir. *ASCE J. Soil Mech. Foundn Engng* 93(4), 551–7.

Jennings, J. E. and K. Knight 1957. The additional settlement of foundations due to collapse of structure on wetting. *Proc. 4th Int. Conf. on soil mechanics and foundation engineering*, London, vol. 1, 316–19.

Jenkins, J. D. and D. E. Bankofier 1972. Hills Creek Dam seepage convection. *Proc. ASCE Spec. Conf. on performance of earth and earth support structures*, vol. 1, 723–33. New York: Am. Soc. Civ. Engrs.

Johnston, G. H. 1969. Dykes on permafrost, Kelsey generating station. *Can. Geotech. J.* 6(2), 139–58.

Johnston, S. J. 1970. Foundation precompression with vertical sand drains. *ASCE J. Soil Mech. Foundn Engng* (1), 145–75.

Jones, D. E. and W. G. Holtz 1973. Expansive soils – the hidden disaster. *ASCE Civ. Engng* August, 87–9.

Jones, J. J. 1967. Deep cutoffs in pervious alluvium, combining slurry trenches and grouting. *Proc. 9th Int. Congr. on large dams*, Istanbul. Paris: International Commission on Large Dams.

Justo, J. L. 1973. The cracking of earth and rockfill dams. *Proc. 11th Int. Congr. on large dams*, Madrid, vol. 4, 921–45. Paris: International Commission on Large Dams.

Kays, W. B. 1977. *Construction of linings for reservoirs, tanks, and pollution control facilities*. New York: Wiley.

Kennard, J. and M. F. Kennard 1962. Selset Reservoir; design and construction. *Proc. Instn Civ. Engrs* 21, 277–304.

Kennedy, J. B., J. T. Laba and M. A. Mossaad 1980. Reinforced earth retaining walls under strip load. *Can. Geotech. J.* 17(3), 382–94.

Kenney, T. C. 1965. Causes of the Vajont Reservoir disaster. *ASCE Civ. Engng* September.

Kenney, T. C. 1967a. The influence of mineral composition on the residual strength of natural soils. *Proc. Geotechnical Conf.*, Oslo, vol. 1, 123–9. Oslo: Norwegian Geotechnical Institute.

Kenney, T. C. 1967b. Slide behaviour and shear resistance of a quick clay determined from a study of the landslide at Selnes, Norway. *Proc. Geotechnical Conf.*, Oslo, vol. 1, 57–64. Oslo: Norwegian Geotechnical Institute.

Kiersch, G. A. 1964. Vajont Reservoir disaster. *Civ. Engng* 34, March.

Kisch, M. 1959. The theory of seepage from clay-blanketed reservoirs. *Geotechnique* 9, 22–8.

Klohn, E. J., C. H. Maartman, R. C. Y. Lo and W. D. L. Finn 1978. Simplified seismic analysis for tailings dams. *ASCE Spec. Conf. on Earthquake Engineering*, Pasadena, CA, vol. 1, 540–56. New York: Am. Soc. Civ. Engrs.

Knight, R. G. 1938. The subsidence of a rockfill dam and remedial measures employed at Eildon Reservoir, Australia. *J. Instn Civ. Engrs*, March.

Koerner, R. M., A. E. Lord and W. M. McCabe 1978. Acoustic emission monitoring of soil stability. *ASCE J. Geotech. Engng* 104(5), 571–82.

Krahn, J., V. E. Price and N. R. Morgenstern 1971. *Slope stability computer program for*

Morgenstern–Price method of analysis. Users Manual 14, University of Alberta, Edmonton, Canada.

Krsmanovic, D. and M. Popovic 1966. Large scale field tests on the shear strength of limestone. *Proc. 1st Congr. of Int. Soc. Rock Mechanics*, Lisbon, vol. 1, 773–80. Lda-Lisbora, Portugal: Bertrand.

Kulhawy, F. H. and J. M. Duncan 1972. Stresses and movement in Oroville Dam. *ASCE J. Soil Mech. Foundn Engng* **98**(7), 653–65.

Kwan, D. 1970. Observations of the failure of a vertical cut in a clay at Welland, Ontario. *Can. Geotech. J.* **8**(2), 283–98.

Lacasse, S. M., C. C. Ladd and A. K. Barsvary 1977. Undrained behaviour of embankments on New Liskeard varved clay. *Can. Geotech. J.* **14**(3), 367–88.

Ladanyi, B. and G. Archambault 1970. Simulation of shear behaviour of a jointed rock mass. *Rock mechanics – theory and practice*, 105–25. New York: American Institute of Mechanical Engineers.

Ladanyi, B., J. P. Morin and C. Pelchat 1968. Post-peak behaviour of sensitive clays in undrained shear. *Can. Geotech. J.* **5**, 59–68.

Ladd, C. C. 1972. Test embankment on sensitive clay. *ASCE Spec. Conf. on earth structures*, Purdue, Lafayette, IN, vol. 1, 101–28. New York: Am. Soc. Civ. Engrs.

Ladd, C. C., J. J. Rixner and D. C. Gifford 1972. Performance of embankments with sand drains on sensitive clay. *ASCE Spec. Conf. on earth structures*, Purdue, Lafayette, IN, vol. 1, 211–42. New York: Am. Soc. Civ. Engrs.

LaFleur, J. and G. Lefebvre 1980. Groundwater regime associated with slope stability in Champlain Clay deposits. *Can. Geotech. J.* **17**(1), 44–53.

Lambe, T. W. 1967. Stress path method. *ASCE J. Soil Mech. Foundn Engng* **93**(6), 309–31.

Lane, K. S. and P. E. Wohlt 1961. Performance of sheet piling and blankets for sealing Missouri River reservoirs. *Proc. 7th Int. Congr. on large dams*, Rome, vol. 4, 255–79. Paris: International Commission on Large Dams.

Larew, H. G. and G. A. Leonards 1962. A repeated load strength criterion. *Proc. Highways Res. Board* **41**, 529–56. Washington, DC: National Academy of Sciences.

LaRochelle, P., G. Lefebvre and P. N. Bilodeau 1977. The stabilization of Saint-Jerome, Lac Saint-Jean. *Can. Geotech. J.* **14**(3), 340–3.

Lauffer, H., E. Neuhauser and W. Schober 1967. Uplift responsible for slope movements during the filling of Gepatsch Reservoir. *Proc. 9th Int. Congr. on large dams*, Istanbul, vol. 1, 669–94. Paris: International Commission on Large Dams.

Law, K. T. and M. Bozozuk 1979. A method of estimating excess pore pressures beneath embankments on sensitive clays. *Can. Geotech. J.* **16**(4), 691–702.

Law, K. T. and P. Lumb 1978. A limit equilibrium analysis of progressive failure in the stability of slopes. *Can. Geotech. J.* **15**(1), 113–22.

Lee, K. L., B. D. Adams and T. M. Vagneron 1973. Reinforced earth retaining walls. *ASCE J. Soil Mech. Foundn Engng* **99**(10), 745–64.

Leps, I. M. 1972. Analysis of failure of Baldwin Hills Reservoir. *Proc. ASCE Spec. Conf. on performance of earth structures*, Purdue University, Indiana, vol. 1, 507–50. New York: Am. Soc. Civ. Engrs.

Leroueil, S., F. Tavenas, B. Trak, P. LaRochelle and M. Roy 1978. Construction pore pressures in clay foundations under embankments. Pt 1: The Saint-Albans test fills. *Can. Geotech. J.* **15**, 54–65.

Lo, K. Y. 1965. Stability of slopes in anisotropic soils. *ASCE J. Soil Mech. Foundn Engng* **91**(4), 85–106.

Lo, K. Y. 1970. The operational strength of fissured clays. *Geotechnique* **21**, 57–74.

Lo, K. Y. and C. F. Lee 1973. Stress analysis and slope stability in strain softening materials. *Geotechnique* **23**(1), 1–12.

REFERENCES

Lo, K. Y. and J. P. Morin 1972. Strength and anisotropy and time effects of two sensitive clays. *Can. Geotech. J.* **9**(3), 261–77.

Lo, K. Y. and A. G. Stermac 1965. Failure of an embankment founded on varved clay. *Can. Geotech. J.* **2**, 234.

McClain, W. C. 1964. Time-dependent behavior of pillars in the Alsace potash mines. *Proc. 6th Symp. on rock mechanics*, Rolla, 489–500. Rolla: University of Missouri.

McConnell, R. G. and R. W. Brock 1904. *Report on the great landslide at Frank, Alberta, Canada*. Dept of the Interior Annu. Rep. 1904, pt 8.

MacDonald, D. H., R. A. Pillman and H. R. Hopper 1960. *Kelsey generating station dam and dykes*. Preprint, 74th Annu. Mtg, Engineering Institute of Canada. Winnipeg: Engineering Institute of Canada.

McGown, A. and A. M. Radwan 1976. The influence of systematic fissuring on the stability of slopes in glacial till. *Proc. 6th Eur. Conf. on soil mechanics and foundation engineering*, Vienna, 67–70. Vienna: Gistel.

McRoberts, E. C. and N. R. Morgenstern 1974. Stability of slopes in frozen soil, Mackenzie Valley, NWT. *Can. Geotech. J.* **11**(4), 554–73.

McRostie, G. C., K. N. Burn and R. J. Mitchell 1972. The performance of tie-back sheet piling in clay. *Can. Geotech. J.* **9**(2), 206–18.

Mansur, C. I. and R. J. Kaufman 1962. *Dewatering, foundation engineering*. New York: McGraw Hill.

Marcuson, W. F., R. F. Ballard and R. H. Ledbetter 1979. Liquefaction failure of tailings dams resulting from the near Izu Oshima earthquake. *Proc. 6th Pan. Am. Conf. on Soil Mech. and Foundn Engng (SMFE)*, Lima, Peru, vol. 2, 69–80. New York: Am. Soc. Civ. Engrs.

Marsal, R. J. 1967. Large-scale testing of rockfill materials. *ASCE J. Soil Mech. Foundn Engng* **93**(2), 27–43.

Marsal, R. J. and W. Pohlenz 1972. The failure of Laguna Dam. *ASCE Spec. Conf. on earth structures*, Purdue, Lafayette, IN, vol. 1, 489–506. New York: Am. Soc. Civ. Engrs.

Marsal, R. J. and D. Resendiz 1971. Effectiveness of cutoffs in earth foundations and abutments of dams. *Proc. 4th Pan Am. Conf. on Soil Mech. and Foundn Engng (SMFE)*, Puerto Rico, 273–312. New York: Am. Soc. Civ. Engrs.

Marsland, A. 1971. The shear strength of stiff fissured clays. In *Stress–strain behaviour of soils*, Proc. Roscoe Memorial Symp., 59–68. Oxfordshire: Foulis.

Mehta, R. 1973. A minor modification in Bishop's expression on the stability of slopes. *Indian Geotech. J.* **3**(1), 56–8.

Menard, L. and Y. Broise 1975. Theoretical and practical aspects of dynamic consolidation. *Geotechnique* **25**, 3–18.

Mesri, G. 1975. Discussion of a new design procedure for stability of soft clays. *ASCE J. Geotech. Engng* **101**(4), 409–12.

Middlebrooks, T. A. 1942. Fort Peck slide. *Trans. Am. Soc. Civ. Engrs* 723–64.

Middlebrooks, T. A. 1977. *Earth dam practice in the United States*, 239–64. Award Winning ASCE papers in Geotechnical Engineering. New York: Am. Soc. Civ. Engrs.

Milbury, F. H. and J. M. Duncan 1972. Stresses and movements in Oroville Dam. *ASCE J. Soil Mech. Foundn Engng* **98**(7), 653–65.

Milne, W. G. and G. C. Rogers 1972. Evaluation of earthquake risk in Canada. *Proc. Int. Conf. on microzonation*, Seattle, WA, vol. 1, 217–30. Washington, DC: National Academy of Sciences.

Mitchell, J. K. and W. N. Houston 1969. Causes of clay sensitivity. *ASCE J. Soil Mech. Foundn Engng* **95**(3), 845–71.

Mitchell, J. K., J. A. Greenberg and P. A. Witherspoon 1973. Chemico-osmosis in fine-grained soils. *ASCE J. Soil Mech. Foundn Engng* **99**(4), 307–21.

Mitchell, J. K., D. R. Hooper and R. G. Campanella 1965. Permeability of compacted clay. *ASCE J. Soil Mech. Foundn Engng* **91**(4), 41–65.

Mitchell, R. J. 1975. Strength parameters for permanent slopes in Champlain Sea Clays. *Can. Geotech. J.* **12**(4), 447–55.

Mitchell, R. J. 1978. *Earthflow terrain evaluation in Ontario.* Res. Rep. 213, Downsview, Ontario: Ministry of Transportation and Communications.

Mitchell, R. J. and W. J. Eden 1972. Measured movements of clay slopes in the Ottawa area. *Can. J. Earth Sci.* **9**, 1001–13.

Mitchell, R. J. and J. A. Hull 1974. Stability and bearing capacity of bottom sediments. *Proc. 14th Int. coastal engineering conf.*, vol. 2, 1252–73. New York: Am. Soc. Civ. Engrs.

Mitchell, R. J. and R. D. King 1977. Cyclic loading of an Ottawa area Champlain Sea Clay. *Can. Geotech. J.* **14**(1), 52–63.

Mitchell, R. J. and M. A. Klugman 1979. Mass instabilities in sensitive Canadian soils. *Engng Geol.* **14**, 109–34.

Mitchell, R. J. and A. R. Markell 1974. Flowsliding in sensitive soils. *Can. Geotech. J.* **11**(1), 11–31.

Mitchell, R. J. and J. D. Smith 1979. Mine backfill design and testing. *Can. Inst. Mining Metall. Bull.* **72**(801), 82–8.

Mitchell, R. J. and J. D. Smith 1981. The percolation–compression test for mine tailings backfill. *Can. Inst. Min. Bull.* **74**(833), 85–9.

Mitchell, R. J. and D. R. Williams 1981. Induced failure of an instrumented clay slope. *Proc. 10th Int. Conf. on soil mechanics and foundation engineering*, Stockholm, vol. 3, 479–84.

Mitchell, R. J. and B. C. Wong 1982. Behaviour of cemented tailings sands. *Can. Geotech. J.* **19**(3), 289–95.

Mitchell, R. J. and P. K. K. Wong 1973. The generalized failure of an Ottawa Valley Champlain Sea Clay. *Can. Geotech. J.* **10**(4), 607–16.

Mitchell, R. J., R. S. Olsen and J. D. Smith 1982. Model studies on cemented tailings used in mine backfill. *Can. Geotech. J.* **19**(1), 14–28.

Mitchell, R. J., D. A. Sangrey and G. S. Webb 1972a. Foundations on the crest of sensitive clay deposits. *ASCE Spec. Conf. on earth structures*, Purdue, Lafayette, IN, vol. 1, 1051–72. New York: Am. Soc. Civ. Engrs.

Mitchell, R. J., J. D. Smith and D. J. Libby 1975. Bulkhead pressures due to cemented hydraulic mine backfills. *Can. Geotech. J.* **12**(3), 362–71.

Mitchell, R. J., K. K. Tsui and D. A. Sangrey 1972b. Failure of submarine slopes under wave action. *Proc. 14th Int. coastal engineering conf.*, Vancouver, vol. 2, 1515–40. New York: Am. Soc. Civ. Engrs.

Mizukoshi, T. and S. Mimura 1975. Studies on the earthquake force in the design of Takase Dam. In *Numerical analysis of dams*, O. J. Naylor, K. G. Stagg and O. C. Zirenkirewcz (eds), Proc. Int. Symp., Swansea, 665–84. University College of Wales, Swansea.

Mollard, J. D. 1973. *Landforms and surface materials of Canada – a stereoscopic atlas and glossary*, 3rd edn. Regina, Sask.: Mollard.

Moorhouse, D. C. and G. L. Baker 1968. Sand densification by heavy vibratory compactor. *ASCE Spec. Conf. on placement and improvement of soil to support structures*, Cambridge, MA, 379–89. New York: Am. Soc. Civ. Engrs.

Mooser, F. 1964. A case of exceptional permeability at El Bosque Dam, Mexico. *Proc. 8th Int. Congr. on large dams*, Edinburgh. Paris: International Commission on Large Dams.

Morgenstern, N. R. 1963. Stability charts for earth slopes during rapid drawdown. *Geotechnique* **13**, 121–31.

Morgenstern, N. R. and V. E. Price 1965. The analysis of the stability of general slip surfaces. *Geotechnique* **15**(1), 79–93.

Morgenstern, N. R. and D. A. Sangrey 1978. Methods of stability analysis. In *Landslides: analysis and control*, 155–69. Spec. Rep. 176, Transportation Research Board, National Academy of Sciences, Washington, DC.

MTC 1976. *Tunnelling technology: an appraisal of the state-of-the-art for application to transit systems.* Ontario Ministry of Transportation and Communications, Downsview.

255

Muller, L. 1964a. Application of rock mechanics in the design of rock slopes. In *State of stress in the Earth's crust*, 575–98. New York: Elsevier.

Muller, L. 1964b. The rock slide in the Vajont Valley. *Rock Mech. Engng Geol.* **11**, 3–4.

Naylor, D. J. 1975. Numerical models for clay core dams. In *Numerical analysis of dams*, O. J. Naylor, K. G. Stagg and O. C. Zienkiewicz (eds), Proc. Int. Symp., Swansea, 489–514. University College of Wales, Swansea.

Newmark, N. M. (1965). Fifth Rankine Lecture: Effects of earthquakes on dams and embankments. *Geotechnique*, **15**(2), 139–59.

Nilsson, T. and B. Loftquist 1955. An earth and rockfill dam on stratified soils. The wet fill method. *Proc. 5th Int. Congr. on large dams*, Paris, vol. 1, 403–13. Paris: International Commission on Large Dams.

O'Connor, M. J. and R. J. Mitchell 1977. An extension of the Bishop and Morgenstern slope stability charts. *Can. Geotech. J.* **14**(1), 144–55.

Odenstad, S. 1951. The landslide at Sköttorp on the Lidan River. *Proc. R. Swed. Geotech. Inst.* **4**, 1–38.

O'Rourke, J. E. 1974. Performance instrumentation in Oroville Dam. *ASCE J. Geotech. Engng* **100**(2), 157–74.

Palladino, D. J. and R. B. Peck 1972. Slope failures in an overconsolidated clay, Seattle, Wash. *Geotechnique* **22**, 563–95.

Palmer, J. H. L. and D. J. Belshaw 1980. Deformations and pore pressures in the vicinity of a precast, segmented, concrete-lined tunnel in clay. *Can. Geotech. J.* **17**(2), 174–84.

Peck, R. B. 1969a. Deep excavations and tunnelling in soft ground. *Proc. 7th Int. Conf. on soil mechanics and foundation engineering*, Mexico, State-of-the-art volume. Mexico: Sociedad Mexicana de Mecánica de Suelos.

Peck, R. B. 1969b. Ninth Rankine Lecture: Advantages and disadvantages of the observational method in applied soil mechanics. *Geotechnique* **19**(2), 171–87.

Peck, R. G., D. U. Deere, J. E. Monsees, H. W. Parker and B. Schmidt 1969. Some design considerations in the selection of underground supports. US Dept of Transport. Springfield, VA: National Technical Information Service.

Pilot, G. 1972. Study of five embankment failures on soft soils. *ASCE Spec. Conf. on earth structures*, Purdue, Lafayette, IN, vol. 1, 81–100. New York: Am. Soc. Civ. Engrs.

Pinkerton, I. L. and A. D. McConnell 1964. Behaviour of Tooma Dam. *Proc. 8th Int. Congr. on large dams*, Edinburgh, vol. 11, 351. Paris: International Commission on Large Dams.

Piteau, D. R. 1972. Engineering geology aspects relating to preliminary damsite investigation on the Nelson River, Manitoba. *Can. Geotech. J.* **9**, 304–12.

Piteau, D. R. and F. L. Peckover 1978. Engineering of rock slopes. In *Landslides: analysis and control*, 193–225. Spec. Rep. 176, Transportation Research Board, National Academy of Sciences, Washington, DC.

Pravdivets, Y. P. and S. M. Slissky 1981. Passing flood waters over embankment dams. *Water Power Dam Constr.* **33**(7; July), 30–3.

Raymond, G. P. 1969. Construction method and stability of embankments on muskeg. *Can. Geotech. J.* **6**, 81–92.

Raymond, G. P. 1972. The Kars (Ontario) embankment foundation. *ASCE Spec. Conf. on earth structures*, Purdue, Lafayette, IN, vol. 1, 319–40. New York: Am. Soc. Civ. Engrs.

Raymond, G. P. 1973. *Foundation failure of New Liskeard embankment*. Highway Research Record no. 463, 1–27, National Academy of Sciences, Washington, DC.

Reuss, R. F. and J. W. Schattenberg 1972. Internal piping and shear deformation of Victor Braunig Dam. *ASCE Spec. Conf. on earth structures*, Purdue, Lafayette, IN, vol. 1, 627–52. New York: Am. Soc. Civ. Engrs.

Richards, B. G. and C. Y. Chan 1969. Predictions of pore water pressures in earth dams. *Proc. 7th Int. Conf. on soil mechanics and foundation engineering*, Mexico, 355–62. Mexico: Sociedad Mexicana de Mecánica de Suelos.

Richardson, G. N. and K. L. Lee 1975. Seismic design of reinforced earth walls. *ASCE J. Geotech. Engng* **101**(2), 167–88.

Robinsky, E. I. 1975. Thickened discharge – a new approach to tailings disposal. *Can. Inst. Min. Bull.*, December, 47–59.

Robinsky, E. I. 1978. Tailings disposal by the thickened discharge method for improved economy and environmental control. *Proc. 2nd Int. tailings symp.*, Denver, CO, vol. 2, 75–95. New York: Am. Soc. Civ. Engrs.

Roscoe, K. H. 1953. An apparatus for the application of simple shear to soil samples. *Proc. 3rd Int. Conf. on soil mechanics and foundation engineering*, vol. 1, 186–91. Zurich: Imprimerie Berichthaus.

Roscoe, K. H., R. H. Bassett and E. R. Cole 1967. Principal axes observed during simple shear of a sand. *Proc. Geotechnical Conf.*, Oslo, vol. 1, 231–7. Oslo: Norwegian Geotechnical Institute.

Roscoe, K. H., A. N. Schofield and C. P. Wroth 1958. On the yielding of soils. *Geotechnique* **8**(1), 22–52.

Royster, D. L. 1980. Landslide remedial measures. *Am. Soc. Engng Geol. Bull.* **16**(2), 290–335.

Rutledge, P. C. and J. P. Gould 1973. Movements of articulated conduits under earth dams on compressible foundations. In *Embankment–dam engineering – Casagrande volume*, R. C. Hirschfeld and S. J. Poulos (eds), 209–38. New York: Wiley.

Samson, L. and P. La Rochelle 1972. Design and performance of an expressway constructed over peat by preloading. *Can. Geotech. J.* **9**(4), 447–66.

Sauer, E. K. 1978. The engineering significance of glacier ice-thrusting. *Can. Geotech. J.* **15**(4), 457–72.

Schmidt, B. 1974. Predictions of settlements due to tunnelling in soil: three case histories. *Rapid excavation and tunnelling Conf.*, San Francisco, CA. New York: Am. Soc. Civ. Engrs.

Schofield, A. N. 1978. Use of centrifugal model testing to assess slope stability. *Can. Geotech. J.* **15**(1), 14–31.

Schofield, A. N. 1981. Twentieth Rankine Lecture: Cambridge geotechnical centrifuge operations. *Geotechnique* **30**(3), 225–68.

Schnabel, P. B., J. Lysmer and H. B. Seed 1972. *SHAKE – a computer program for earthquake response analysis of horizontally layered sites*. Rep. EERC-72-12, University of California, Berkeley.

Scott, J. S. and E. W. Brooker 1968. *Geological and engineering aspects of Upper Cretaceous shales in Western Canada*. Geological Survey of Canada, Paper 66-37. Ottawa: Department of Energy, Mines and Resources.

Seed, H. B. 1966. A method for earthquake resistant design of earth dams. *ASCE J. Soil Mech. Foundn Engng* **92**(1), 13–41.

Seed, H. B. 1967. Earthquake-resistant design of earth dams. *Can. Geotech. J.* **4**(1), 1–27.

Seed, H. B. 1968. Landslides during earthquakes due to soil liquefaction. *ASCE J. Soil Mech. Foundn Engng* **94**(5), 1055–122.

Seed, H. B. and C. K. Chan 1959. Structure and strength characteristics of compacted clay. *ASCE J. Soil Mech. Foundn Engng* **85**(5), 87–128.

Seed, H. B. and C. K. Chan 1966. Clay strength under earthquake loading conditions. *ASCE J. Soil Mech. Foundn Engng* **92**(2), 53–78.

Seed, H. B. and G. R. Martin 1966. The seismic coefficient in earth dam design. *ASCE J. Soil Mech. Foundn Engng* **92**(3), 25–58.

257

Seed, H. B. and W. H. Peacock 1971. Test procedures for measuring soil liquefaction characteristics. *ASCE J. Soil Mech. Foundn Engng* **97**(8), 1099–119.

Seed, H. B. and H. A. Sultan 1967. Stability analyses for a sloping core embankment. *ASCE J. Soil Mech. Foundn Engng* **93**(4), 69–80.

Seed, H. B. and R. V. Whitman 1970. Design of earth retaining structures for dynamic loads. *ASCE Spec. Conf. on Earth-retaining Structures*, Cornell University, Ithaca, NY, 103–48. New York: Am. Soc. Civ. Engrs.

Seed, H. B. and S. D. Wilson 1967. The Turnagain Heights landslide, Ankorage, Alaska. *ASCE J. Soil Mech. Foundn Engng* **93**(4), 325–53.

Seed, H. B., J. M. Duncan and I. M. Idriss 1975a. Criterion and methods for static and dynamic analysis of earth dams. In *Numerical analysis of dams*, O. J. Naylor, K. G. Stagg and O. C. Zienkiewicz (eds), Proc. Int. Symp., Swansea, 563–88. University College of Wales, Swansea.

Seed, H. B., K. L. Lee and I. M. Idriss 1969. Analysis of Sheffield Dam failure. *ASCE J. Soil Mech. Foundn Engng* **95**(6), 1453–90.

Seed, H. B., K. L. Lee, I. M. Idriss and F. Makdisi 1975b. Slides in the San Fernando Dams during the earthquake of February 9, 1971. *ASCE J. Geotech. Engng* **101**(GT7), 651–90.

Seychuk, J. L. and L. R. Lahti 1979. Groundwater control to the rescue: two case histories. *Can. Geotech. J.* **16**(4), 716–33.

Shields, D. H. 1975. Innovations in tailings disposal. *Can. Geotech. J.* **12**(3), 320–5.

Simonds, A. W. 1977. *Final foundation treatment at Hoover Dam*, 33–55. Award Winning ASCE Papers in Geotechnical Engineering.

Singh, R., D. J. Henkel and D. A. Sangrey 1973. Shear and Ko swelling of overconsolidated clay. *Proc. 9th Int. Conf. on soil mechanics and foundation engineering*, Moscow, vol. 1, 367–76. Printed in the USSR under authority of the USSR National Society of Soil Mechanics and Foundation Engineering.

Skempton, A. W. 1954. The pore pressure coefficients A and B. *Geotechnique* **4**(4), 143–7.

Skempton, A. W. 1957. Discussion on the planning and design of the new Hong Kong Airport. *Proc. Instn Civ. Engrs* **7**, 305–7.

Skempton, A. W. 1964. Long-term stability of clay slopes. *Geotechnique* **14**(2), 77–101.

Skempton, A. W. 1970. First-time slides in overconsolidated clays. *Geotechnique* **20**(3), 320–4.

Skempton, A. W. and L. Bjerrum 1957. A contribution to the settlement analysis of foundations on clays. *Geotechnique* **7**, 168–78.

Skempton, A. W., R. L. Schuster and D. J. Petley 1969. Joints and fissures in London Clay at Wraysbury and Edgware. *Geotechnique* **19**(2), 205–17.

Smith, J. D. and R. J. Mitchell 1982. Design and control of large hydraulic backfill pours. *Can. Inst. Min. Bull.* **75**(838), 102–11.

Sowers, G. F. 1962. *Earth and rockfill dam engineering*. London: Asia Publishing House.

Sowers, G. F. and D. L. Royster 1978. Field investigation. In *Landslides: analysis and control*, 81–111. Spec. Rep. 176, Transportation Research Board, National Academy of Sciences, Washington, DC.

Stafford, D. B. and L. J. Langfelder 1971. Airphoto survey of coastal erosion. *Photogramm. Engng* **37**, 565–75.

Sutcliffe, F. H. 1965. A till cofferdam in the St Lawrence River. *Can. Geotech. J.* **2**, 261–70.

Sverdrup, H. and W. H. Munk 1946. Wind waves and swell principles in forecasting. US Hydrol Surv. Misc, Publ. 11275, Washington, DC.

Tavenas, F. 1979. The behaviour of embankments on clay foundations. *Proc. 32nd Canadian Geotechnical Conf.*, Quebec City, State-of-the-art volume, 40.

Tavenas, F., J.-Y. Chagnon and P. LaRochelle 1971. The Saint-Jean-Vianney landslide: observations and eye-witness accounts. *Can. Geotech. J.* **8**(3), 463–78.

Tavenas, F., R. Garneau, R. Blanchet and S. Leroueil 1978. The stability of stage-constructed embankments on soft clays. *Can. Geotech. J.* **15**(2), 283–305.
Taylor, D. W. 1937. Stability of earth slopes. *J. Boston Soc. Civ. Engrs* **24**(3), 197–246.
Terzaghi, K. 1943. *Theoretical soil mechanics*. New York: Wiley.
Terzaghi, K. 1956. Varieties of submarine slope failures. *Proc. 8th Texas soil mechanics and foundation engineering conf.* New York: Am. Soc. Civ. Engrs.
Terzaghi, K. and R. B. Peck 1967. *Soil mechanics in engineering practice*, 2nd edn. New York: Wiley.
Thomas, E. G., J. Nantel and K. R. Notley 1979. *Fill technology in underground metalliferous mine*. Kingston, Ont.: International Academic Services.
Thomas, H. H. 1976. *The engineering of large dams*, pt 1. New York: Wiley.
Trow. W. A. 1974. Temporary and permanent earth anchors: three monitored installations. *Can. Geotech. J.* **11**(2), 257–68.
Turnbull, W. J. and C. I. Mansur 1961. Investigation of underseepage, Missippi River levees. *ASCE Trans.* **126**, 1429–81.

USCE (US Corps of Engineers) 1971. *Earth and rockfill dams, general design and construction considerations*. Rep. EM 1110-2-2300, USCE, Vicksburg, MS.
USDI (US Dept of the Interior) 1977. *Teton dam failure: a report of findings*. Washington, DC: US Govt Printing Office.

VanDine, D. F. 1980. *Engineering geology and geotechnical study of Drynoch landslide, BC*. Paper 79-31, Energy, Mines and Resources, Canada.
Varnes, D. J. 1978. Slope movement and types and processes. In *Landslides: analysis and control*. Spec. Rep. 176, Transportation Research Board, National Academy of Sciences, Washington, DC.
Vidal, H. 1966. La terre armee. *Ann. Inst. Technique Bâtim. Travaux Publics*, nos 223–229, July–August 888–938.
Vidal, H. 1969. *The principle of reinforced earth*. Highway Res. Record no. 282, National Academy of Sciences, Washington, DC.

Wade, N. H. and D. J. Henkel 1966. Plane strain tests on a saturated remoulded clay. *ASCE J. Soil Mech. Foundn Engng* **92**(6), 67–80.
Wafa, T. A. and A. H. Labib 1967. The great grout curtain under the High Aswan. *Proc. 9th Int. Congr. on large dams*, Istanbul. Paris: International Commission on Large Dams.
Wallways, M. 1964. Soil compaction in depth by means of piles, made with compacted sand or granulated slag. *Technique Travaux*, November–December.
Walters, R. C. S. 1971. *Dam geology*, 2nd edn. London: Butterworth.
Ward, W. H., A. Marsland and S. G. Samuels 1965. Properties of London Clay at the Ashford Common shaft: *in situ* and undrained strength tests. *Geotechnique* **15**, 321–44.
Watanabe, H. 1975. A numerical method of seismic analysis for rock and earth fill dams and verification of its reliability through both model test and observation of earthquake on an actual dam. In *Numerical analysis of dams*, O. J. Naylor, K. G. Stagg and O. C. Zienkiewicz (eds), Proc. Int. Symp., Swansea, 745–66. University College of Wales, Swansea.
Way, D. S. 1973. *Terrain analysis*. Stroudsburg, PA: Dowden, Hutchinson & Ross.
Wayment, W. R. 1978. Backfilling with tailings – a new approach. In *Mining with backfill*, Proc. 12th Can. Rock Mech. Symp., Sudbury, 111–16. Can. Inst. Mining Metall. Spec. Vol. 19, Montreal: Can. Inst. Mining Metall.
Weaver, W. S. and R. Luka 1970. Laboratory studies of cement-stabilized mine tailings. *Can. Inst. Min. Bull.*, September, 988–1001.
Weber, W. G., Jr 1962. Construction of a fill by a mud displacement method. *Proc. Highways Res. Bd* **41**, 591–610.

Webster, J. L. 1970. Mica Dam designed with special attention to control of cracking. *Proc. 10th Int. Congr. on large dams*, Montreal, vol. 1, 487–507. Paris: International Commission on Large Dams.

Whitham, K., W. G. Milne and W. E. T. Smith 1970. *The new seismic zoning map for Canada.* Ottawa: Canadian Underwriters.

Wilkes, P. F. 1972. An induced failure at a trial embankment at King's Lynn, Norfolk, England. *ASCE Spec. Conf. on earth structures*, Purdue, Lafayette, IN, vol. 1, 29–64. New York: Am. Soc. Civ. Engrs.

Williams, D. R., P. M. Romeril and R. J. Mitchell 1979. Riverbank erosion and recession in the Ottawa area. *Can. Geotech. J.* **16**, 641–50.

Wilson, S. D. and P. E. Mikkelsen 1978. Field instrumentation. In *Landslides: analysis and control*, 112–37. Spec. Rep. 176, Transportation Research Board, National Academy of Sciences, Washington, DC.

Wilson, S. D. and R. Squier 1969. Earth and rockfill dams. *Proc. 7th Int. Conf. on soil mechanics and foundation engineering*, Mexico, State-of-the-art volume, 137–220. Mexico: Sociedad Mexicana de Mecánica de Suelos.

Wolfskill, L. A. and T. W. Lambe 1967. Slide in the Siburua Dam. *ASCE J. Soil Mech. Foundn Engng* **93**(4), 107–33.

Wong, P. K. K. and R. J. Mitchell 1975. Yielding and plastic flow of sensitive cemented clay. *Geotechnique* **15**(4), 763–82.

Wright, S. C. and R. S. Dunham 1972. Bottom stability under wave induced loading. *Offshore Technical Conf.*, Texas, paper OTC 1603. New York: Am. Soc. Civ. Engrs.

Wu, T. H. 1976. *Soil mechanics*, 2nd edn. Boston: Allyn and Bacon.

Wu, T. H. and L. M. Kraft 1970. Safety analysis of slopes. *ASCE J. Soil Mech. Foundn Engng* **96**(2), 609–30.

Wu, T. H. and D. A. Sangrey 1978. Strength properties and their measurement. In *Landslides: analysis and control*, 139–52. Spec. Rep. 176, Transportation Research Board, National Academy of Sciences, Washington, DC.

Yong, R. N. amd V. Silvestri 1979. Anisotropic behaviour of a sensitive clay. *Can. Geotech. J.* **16**(2), 335–50.

Yong, R. N., E. Alonso, M. M. Tabba and P. B. Fransham 1977. Application of risk analysis to the prediction of slope instability. *Can. Geotech. J.* **14**(4), 540–53.

Young, A. R. 1963. West Water Reservoir and Dam. *Civ. Engng Public Works Rev.*, October, 1249–51.

Zaruba, Q. and V. Mencl 1969. *Landslides and their control.* New York: Elsevier.

Answers to problems

Numerical answers are provided where applicable. Some questions require assumptions and discussion type answers – for these questions, short statements are given and relevant sections of the text are referred to.

Section 1.10   Problems on air photo interpretation

1 (a) 18°. (b) Lacustrine or marine (Sections 1.6 & 1.8.1). (c) Marginally stable, 150 m.
2 9.5° to 11.5°, slide evident, clay with sand cover.
3 Contain surface water, organic over sand, 12 m at 1:1 (rock).
4 Earthflows (ancient), retrogressive flowside and translatory (recent).
5 Parallel (Section 1.3.1).
6 (a) Stagnant ice, retreating ice and advancing ice landforms. (b) Pleistocene. (c) Section 1.5, Table 1.1.
7 (a) Outwash. (b) Two older slides (stable). (c) V-gullies, deciduous, sapping.
8 (a) Ice. (b) Ancient earthflow. (c) Compare Figures 1.20 and 1.13. (d) β = 12° (4.8:1). (e) Pollution of shallow wells.

Section 3.3   Problems on bearing capacity and tunnels

1 (a) Final stage after about 30 months (Sections 3.3.1 & 3.1.3). Sand drains needed. (b) Sand drains at 3 m spacing, three stages to 6.5 m, 11 m, 15 m at 4 to 5 month delay.
2 (a) 56 mm with 33 mm at 2.4 m left and right for 8 m depth. 50 mm with 30 mm at 3.2 m left and right for 12 m depth. (b) 200 mm for 8 m depth and 250 mm for 12 m depth (Table 2.9), extending 35 m to 55 m left and right. (c) Piping.
3 7.3 kPa for stability, $z_w \gamma_w$ for piping control if silty layers.

Section 4.8   Problems on slope stability

1 For $c' = 0$, $h_w = 0$, $F = \dfrac{2Cu}{\gamma H}(\cot \alpha + \cot \beta) + \tan \phi'/\tan \alpha$. Equations within $\pm 10\%$ for $20° \le \phi' \le 45°$ for practical values of $Cu/\gamma h$, $\cot \beta$.

2 Shallow circle $F = 6.76c'/\gamma H$, deeper circle $F = 6.37c'/\gamma H$; compare to $N_s = 6.13$ and 5.56 (respectively) from Table 4.1.

3 Retaining wall economical for $t > 1.4$ months.

4 When $c' = 0$, $F = 0.8$, tests should be done on saturated samples (Section 2.1.6).

5 At $r_u = 0.41$, $F = 1.56$ (stable) but potential for retrogression high (see Eqs 4.11 & 4.13). Risk low but inspection recommended.

6 3.5:1 ($F = 1.35$), 10% remedial.

7 (a) $\phi'_r = 7°$, $\gamma_r = 2.5\gamma_w$. (b) Toppling (safe), translatory (safe).

8 Use computer or Hoek and Bray (1974) charts. For pit slope estimate σ'_N (avg) $= \gamma H \cos \beta/2$, giving $\tau_{avg} = 200$ kPa. For dragline need $Cu = c = 215$ kPa.

9 (a) Translatory or composite. (b) 28.5°. (c) 10.5 kPa. (d) Intersect at σ'_N (avg) $= 20$ kPa $= (W \cos \alpha - U)/L$. (e) Rotational. (f) 57.3°. (g) 27 kPa. (h) No, two different circles. (i) North slope, composite.

10 1.6 static, 1.4 earthquake.

11 (a) Section 2.1.9. (b) Westerly strike: north slope $= 35°$, south slope $= 63°$. North strike: east slope $= 70°$, west slope $= 50°$. Intermediate slopes 45° to 56°.

12 $F = 0.7$ when hydrostatically saturated, closing justified.

13 (a) $F = 0.6 + 0.26(\cot \beta - 1)$. (b) $1 + 4.58(\tan \phi' - 0.2)$. (c) 1.1 (little variation). (d) ϕ' high.

14 (a) If $c' > 0$, F will reduce below 1.3 if no cut is made. (b) Upstream groundwater pressure will increase if drainage ditch not provided.

15 Section 4.3.1.

Section 5.9 Problems on earth dams

1 (a) 2:1 downstream, 2.5:1 upstream (excavate organics). (b) Upstream $F = 0.97$ for drawdown and earthquake (may increase to 2.8:1). (c) Major leakage through sand and gravel (about 3×10^{-4} m³ s⁻¹), $u > \sigma_v$ downstream, require relief wells or cutoff through sand and gravel. (d) 3 m, 0.08 tonnes.

2 (a) Wide core. (b) Compaction of 1, 2, 3 zones too high (see Section 5.6). (c) The D_{15} ratio from zone 1 to 3 is too low. (d) Internal settlements are small but distortion due to foundation settlement is 7% (material testing for rupture). (e) $E_h \epsilon_h < 2$ kPa (no risk unless chord length < 50 m). (f) 0.7.

3 (a) 1.5:1 downstream, 2.7:1 upstream for rapid drawdown, check with core in place (Fig. 5.7). $\Delta H/H \cot \beta = 0.05$ (high distortion). (b) Remove silt/sand cover and excavate silty clay (5×10^6 m³), replace silt/sand to elev. 135 m (8.5×10^6 m³), grout curtain required to control seepage, $r_u = 0$ downstream (Fig. 5.10). (c) Section 5.7, Figure 5.27.

4 (a) Surface erosion or differential settlements. (b) Both have overflow channels to depression adjacent downstream pond, drainage to river downstream.

5 All flow through filter with $\Delta h = t$. $N_f/N_d = 2.25h/L$. For average downstream slope angles $N_f/N_d \approx 1$.

Section 6.5 Problems on mine backfill and subsidence

1 40%, see Mitchell and Smith (1979).

2 $V = P/n = (10/0.45) = 22$ cm/h since $i = 1$. $\omega \le 10\%$.

3 $h_c \approx 2$ m, need 1 m of drained material, minimum depth = 3 m.

4 $H = 120$ m, 40 m pillar width, 20 m stope width, cost = $1.15 per tonne of ore. A 3 m thick remnant has a net worth of about 0.01 times the stope cement costs. Unless production requirements override, it is more economical to leave remnant pillars.

5 (a) 4.8×10^{-3} m³ s⁻¹. (b) 0.21 m³ s⁻¹ (7.5×10^6 litre h⁻¹).

6 0.3 m for 10:1 T:C and 0.6 m for no cement. Compression calculation gives about 0.28 m for both conditions.

7 (a) 2×10^{-5} m³ s⁻¹ per metre of slot length. (b) 0.17 m and complete drainage of overlying area, grout confined sand zone to 50 m each side of slot to $k \le 10^{-7}$ m s⁻¹ (Section 5.3.2). (c) Caving occurs, $y_c = 2$ mm to $\delta_m = 0.2$ m.

8 $c'/\gamma H$ and $\tan \phi'$ same in model and prototype. Either measure apparent c' due to capillarity or eliminate by saturation and flow under $i = 1$ during model exposure.

9
Beam depth, m	30	20	10	5	0
Subsidence, cm	0	0.5	5	18	47

Index

accelerometer 203
air photo scales 7
anisotropy 51, 70
arching 195, 222, 233

bearing capacity 88, 91
Bishop simplified solution 113, 128
bulkheads 222

capillary water 62
cement stabilization 226
cement tailings design 227
classification 40
coefficients of
    consolidation 81
    lateral earth pressure 43, 152
    permeability 59, 67
    volume compression 79
compaction 75, 200, 202
compressibility 76, 79
compression index 77
confined modulus 196, 237
consolidation 77, 81
cores and cutoffs 172
creep 54, 80, 143
crest loadings 135
critical centre 125
critical hydraulic gradient 61
critical stress analysis 115
critical void ratio 49
cyclic loading 53

dam
    design 168
    distortion 192
    inspection 204
    shells 168
    site selection 25, 168
degree of saturation 62
de-watering 72, 147
dilative strength 49, 56
direct shear tests 42
displacement construction 95, 211
drainage 62, 72, 94, 147, 223
drawdown 139, 198

earthflow 11, 108, 118
earthquake loadings 135, 198
effective stress 43
erosion protection 148, 176
extensometer 203

factor of safety 87, 109, 111, 112, 128, 143
field tests 41

filter design 175
flight lines 4
flow nets 71
foundation treatment 94, 180
fully softened strength 123

geotextiles 96, 176
glacial landforms 12
grain crushing 45
grouting 100, 186

hydraulic gradient 61

inclinometer 202
instrumentation 202
intermediate principal stress 51

laboratory tests 42
liquefaction 53, 136, 225, 233
liners 212

maximum dry density 76, 77
mine backfill 219
Mohr–Coulomb failure criterion 43

optimum water content 76
organic soils 40, 95
overconsolidation ratio 47
overload factor 97

partly saturated soil 53
percolation rate 223
permeability tests 63, 239
piezometer compliance 64
piezometers 63, 93, 127, 202
piping 61, 87, 174
pore-water pressure 43, 80, 200
    coefficients 46, 199
    ratio 112, 125
    under embankments 92, 200
preconsolidation pressure 77, 92
principal stresses 39, 51
process-water impoundments 212
progressive failure 123, 171

rapid drawdown 135, 198
residual strength 50, 124
retaining walls 151
rip-rap 167, 178
rock
    definition 1
    faults 27
    landforms 12
rotational landsliding 106, 112

sapping 30
seepage control 172
seepage distance 70
settlement calculations 85, 192
shield tunneling 96
site investigation 38, 164
slickenside 121, 124
slope
  monitoring 150, 201
  remedial measures 145
  stability 22, 104, 170
  stability design charts 127
slurry trenching 186
soil definition 1
spillways 168, 179
stability numbers for slopes 134
stage construction 92
stereo photos 4
stress distribution under embankments 84

stress paths 57
stress–strain relations 50
submergence 135
subsidence 24, 99, 219, 236
swelling index 77

tailings dam 205
Terzaghi consolidation theory 80
time effects 54, 82, 122
time factor 81
triaxial tests 42, 57
tunnel stability 97

undrained strength 48, 87, 133

vane strength 89, 133

wedge sliding 107, 117
well pumping tests 68